PRAISE FOR BREAKFAST WITH EINSTEIN

"Physics is everywhere and in every thing, and no one explains physics better than Chad Orzel. This book is a meal for your mind."
—John Scalzi

"Why don't light bulbs fry us with deadly radiation? Why can't you stick your hand through a solid wall? Why isn't every scrap of metal a magnet? So many science books focus on the latest wacky cosmic discovery, but Orzel shows how the ordinary world around us is already plenty weird."
—George Musser, contributing editor at *Scientific American* and author of *Spooky Action at a Distance*

"Common wisdom paints quantum mechanics as one of the most abstract and esoteric of subjects, daunting for non-experts. Yet as Chad Orzel wonderfully shows in *Breakfast with Einstein*, a full gamut of our commonplace daily activities—from boiling water for tea on a glowing range to taking and exchanging photos with our electronic cameras and phones—depends on quantum rules. By focusing on how quantum mechanisms guide the workings of his typical morning routine, Orzel cleverly brings those important principles close to home. A must-read for anyone fascinated with how the quantum revolution explains how things work."
—Paul Halpern, author of *The Quantum Labyrinth: How Richard Feynman and John Wheeler Revolutionized Time and Reality*

"William Blake saw the world in a grain of sand. Chad Orzel sees the universe in a slice of toast. Orzel is a master at bringing abstract ideas like relativity and quantum mechanics down to earth without ever skimping on the science. This fun, engaging, and deeply informative book is definitive proof that everything is fascinating when you look closely enough. I'll never see my breakfast the same way again."
—Amanda Gefter, author of *Trespassing on Einstein's Lawn*

"*Breakfast with Einstein* offers a clear and entertaining introduction to the wonders of quantum mechanics, showing that these principles surround us and are employed regularly in our everyday lives. Chad Orzel is the perfect guide to the world of atoms and photons, demonstrating that even our morning breakfast rituals are not possible without the wonders of modern physics."
—James Kakalios, professor of physics at the University of Minnesota and author of *The Physics of Superheroes* and *The Physics of Everyday Things*

"Orzel draws us in with the everyday experience. And then we find we are on a journey of more than 100 years of physics. The reader is rewarded not only with a deeper understanding of everyday things but also how physicists themselves look at the world every day."
—David Saltzberg, professor of physics and astronomy at UCLA

"Chad Orzel's new book is a masterfully told story about the myriad ways that physics shapes our lives."
—Sabine Hossenfelder, physicist and author of *Lost in Math*

BREAKFAST WITH EINSTEIN

BREAKFAST WITH EINSTEIN

THE EXOTIC PHYSICS OF EVERYDAY OBJECTS

CHAD ORZEL

BenBella Books, Inc.
Dallas, TX

BenBella Books, Inc.
10440 N. Central Expressway, Suite 800
Dallas, TX 75231
www.benbellabooks.com
Send feedback to feedback@benbellabooks.com

Printed in the United States of America
10 9 8 7 6 5 4 3 2 1

Library of Congress Cataloging-in-Publication Data is available upon request.
9781946885357 (trade paper)
9781946885661 (e-book)

Editing by Alexa Stevenson
Copyediting by Scott Calamar
Proofreading by James Fraleigh and Laura Cherkas
Indexing by WordCo Indexing Services, Inc.
Text design and composition by Aaron Edmiston
Cover design by Pete Garceau
Cover photo © iStock / Kativ and Susoy
Printed by Versa Press

Distributed to the trade by Two Rivers Distribution, an Ingram brand
www.tworiversdistribution.com

**Special discounts for bulk sales (minimum of 25 copies) are available.
Please contact Aida Herrera at aida@benbellabooks.com.**

For David, my favorite Little Dude.
Of all the little dudes I know, you're the one I love the best.

CONTENTS

INTRODUCTION

The sun comes up not long before my alarm clock starts beeping, and I get out of bed to start the day. It's still dark in the hallway when I leave the bedroom, the status light on the smoke detector casting a faint light on the wall. Down in the kitchen, I put water on for tea—checking for the glow of the heating element to make sure I haven't groggily put the kettle on the wrong burner again. I open the refrigerator to start breakfast, careful not to dislodge the many works of art held to the door with magnets. I slide a couple of slices of bread into the toaster oven, jiggling the rack when it sticks a bit, and lean against the counter while I wait.

My tea is still a bit too hot to drink, but I savor the aroma of the rising steam as it cools, and start up the computer to see what's going on in the world. My social media feeds are full of the usual overnight fare—morning news from Europe and Africa, evening stories from Asia and Australia, digital photos of the kids and cats of friends around the world. My email is mostly from students requesting homework help, plus a couple of receipts and tracking notices from online purchases.

After my tea and toast, I grab the dog's leash and we head out for our morning walk. When we return, it'll be time to get the kids up and ready for school. Once they're on the bus, I'll head off to school myself, to teach another class of students about the physics all around them.

hen I tell people I'm a physicist, I'm often met with questions about exotic phenomena, drawing on some of the vivid and colorful examples that have emerged from decades of debate about quantum theory. People ask about Schrödinger's famous cat, alive and dead at the same time, or about the quantum entanglement that Einstein derided as "spooky action at a distance," or whether God *really* plays dice with the universe. These topics capture the imagination of nonscientists as well as professional physicists, because they confound our intuition about how the world works.

While physicists and popularizers of physics have been very successful at pushing some of these abstract and odd-seeming ideas into popular culture, in a way we are also victims of that success. Most people who have heard of these strange and captivating phenomena also think of them as the sort of things that only turn up in a multibillion-dollar experiment like the Large Hadron Collider, or in extreme astrophysical environments like near the event horizon of a black hole. The counterintuitive nature of these phenomena and the metaphorical language we must use in order to discuss them in nonmathematical terms combine to convince most people that quantum physics has no relevance at all to everyday life.

It might come as a surprise, then, to learn that nothing in the description of a mundane morning that began this introduction would be possible without "exotic" quantum physics. The time marked by our alarm clocks can be traced back to energy states within atoms that exist because of the wave nature of electrons. The semiconductor chips at the heart of the computers we use to send one another funny cat memes can't be understood without quantum superpositions like that of Schrödinger's infamous zombie cat. Neither the chemistry of aroma nor even the stability of solid matter that keeps our breakfasts from falling through the table can be explained without the strange statistical properties of quantum spins.

On closer inspection, it turns out that our everyday world is profoundly influenced by the "exotic" and "abstract" phenomena of quantum physics. Even the most ordinary of activities, those that make up our morning routine, are fundamentally quantum, when you dig into them a bit.

This might sound unlikely at first, but if you think about it, it *has* to be true. After all, physicists inhabit the same everyday world as everyone else. While state-of-the-art physics experiments use lasers and particle accelerators to probe that world at a level far beyond our everyday experience, even the most complicated experiments and observations must begin and end right here in ordinary reality. And the sophisticated apparatus employed for those experiments has mundane roots: the tools and techniques used to study even the most arcane aspects of physics were built up deliberately over many years, following small clues to ever stranger phenomena. The clues that lead us to the exotic and the abstract began with hints and mysteries in the behavior of ordinary objects. If quantum physics didn't affect the everyday, macroscopic world, we never would've needed to discover it.

The story of that discovery begins with observations and technologies that are very familiar to nearly anyone who has ever made breakfast. The very first quantum theory—in fact, the theory that introduced the word "quantum" to physics—was invented by Max Planck to explain the red glow of a hot object like the heating element in an electric stove or toaster. Quantum ideas were first applied to material objects in Niels Bohr's model of the hydrogen atom; you see the underlying physics in action any time you use a fluorescent light.

The history of quantum physics is also a history of scientists making bold leaps and lucky guesses. Planck and Bohr introduced their quantum models as desperate tricks to explain phenomena that classical physics simply couldn't. Louis de Broglie proposed that electrons might behave like waves for reasons of mathematical elegance, and the wave nature of matter turns out to be essential for understanding and controlling how electric current moves, enabling an enormous range of modern technologies. Wolfgang Pauli explained the conceptual basis of chemistry in a stroke when he introduced his exclusion principle. "Pauli exclusion" also turns out to be crucial to understanding problems he hadn't yet considered, like the physics of refrigerator magnets and why solid objects don't collapse in on themselves.

Albert Einstein was a key player in all of this—his name isn't on the cover just to sell books. We mostly associate Einstein with his theory of relativity, a different (and just as fascinating) branch of modern physics, and if he's mentioned at all in connection with the quantum, it's usually to quote one of his many pithy and disdainful remarks on the theory from his later years.

In fact, though, Einstein played a pivotal role in the development of quantum physics. In 1905, the same year he launched relativity, he also picked up and extended Planck's quantum model to explain the photoelectric effect, the physics of which is essential for the operation of the digital cameras we use to so extensively document our modern lives. A decade later, he elaborated on the interaction between light and atoms in a way that laid the foundation for the invention of lasers, which are the cornerstone of modern telecommunications. And even as he fell away from the mainstream of quantum physics, he made a valuable contribution: his parting shot introduced the idea of entanglement, which is at the heart of many proposals for the next generation of quantum technologies involving unbreakable encryption and computers of unprecedented power.

My goal with this book is to reveal the quantum foundations of everyday reality by digging into the mundane morning described earlier. In the following chapters, I'll explore many of the activities described to show how an ordinary weekday routine depends on some of the weirdest phenomena ever discovered. And as I explain how quantum effects connect with our daily life, I'll also share the story of some of the clues physicists followed to uncover them.

The intent here is not to drag quantum physics down until it is as unremarkable as a weekday breakfast. Rather, I hope to elevate the everyday by showing the wonder and amazement that can be found in even the simplest activities, ones we take for granted. Quantum physics is one of the greatest intellectual triumphs of human civilization, full of mind-expanding and imaginative new ideas. It's also all around us, every day, if we just know where to look.

CHAPTER 1

SUNRISE: THE FUNDAMENTAL INTERACTIONS

The sun comes up *not long before my alarm clock starts beeping, and I get out of bed to start the day . . .*

I t might seem like cheating to start a book on the quantum physics of everyday objects by talking about the sun. After all, the sun is a vast sphere of hot plasma, a bit more than a million times the volume of Earth, floating in space ninety-three million miles from here. It's not an everyday object in the same way as, say, an alarm clock that you can pick up and throw across the room when it wakes you after a too-short sleep.

On the other hand, in a sense the sun is the most important everyday object of all, even beyond the glib observation that a day doesn't start until the sun rises. Without the light we receive from the sun, life on Earth would be utterly impossible—the plants we rely on for food and oxygen wouldn't grow, the oceans would freeze, and so forth. We're dependent on the light and heat of the sun for our entire existence.

For the purposes of this book, the sun is also a useful vehicle for a kind of *dramatis personae*, introducing the key players of quantum physics: the twelve fundamental particles that make up ordinary matter, and the four fundamental interactions between them.

The twelve fundamental particles—particles that cannot be broken down any further into even smaller parts—are divided into two "families," each with six particles. The *quark* family consists of the up, down, strange, charm, top, and bottom quarks, and the *lepton* family contains the electron, muon, and tau particles, along with electron, muon, and tau neutrinos. The four fundamental interactions are gravity, electromagnetism, the strong nuclear interaction, and the weak nuclear interaction. You can often find these particles and interactions enumerated on a colorful chart hanging in a physics classroom, collectively referred to by the sadly generic name "the Standard Model of physics." The Standard Model encapsulates everything we know about quantum physics (and also about the ability of physicists to come up with catchy names), and ranks as one of the greatest intellectual achievements of human civilization.[*] The sun is a perfect introduction to the Standard Model, because all four of the fundamental interactions have a role to play in order for the sun to shine.

So, we'll start our story with the sun, taking a whirlwind tour of its inner workings to illustrate the essential physics that powers everything else we do. We'll go through each of the fundamental interactions in turn, beginning with the best known and most obvious of these forces: gravity.

---

[*]   For a more complete overview of the physics involved in the Standard Model, I recommend Robert Oerter's *The Theory of Almost Everything* (Plume, 2006); the historical development is described in detail in Frank Close's *The Infinity Puzzle* (Basic, 2013).

GRAVITY

If you were to generate sports radio–style "power rankings" of the fundamental interactions of the Standard Model, three of the four have a decent case for claiming the top spot. If pressed to make a choice, though, I'd probably give the honors to gravity, because gravity is ultimately responsible for the existence of stars, and thus for most of the atoms making up our bodies and everything around us, enabling our silly conversations about ranking fundamental forces.

In our everyday lives, gravity is probably the most familiar and inescapable of the fundamental interactions. It's gravity that you fight against when getting out of bed in the morning, and gravity that keeps me from being able to dunk a basketball (well, gravity, and being woefully out of shape . . .). We spend the vast majority of our lives feeling the pull of gravity, which makes its temporary absence—as in amusement park rides featuring sudden drops—fascinating, and even thrilling.

That familiarity also makes gravity one of the most-studied forces in the history of science. People have been thinking about how and why objects fall to the earth for at least as long as we have records of people pondering the workings of the natural world at all. Popular legend traces the origin of physics to a young Isaac Newton being struck (literally, in some versions) by the fall of an apple from a tree, and thus impelled to invent a theory of gravity. Contrary to the image conjured up by this apocryphal tale, though, scientists and philosophers were already well aware of gravity, and had devoted significant thought to how it works. By Newton's day, experiments by Galileo Galilei, Simon Stevin, and others had even made some quantitative headway on the subject, establishing that all dropped objects, regardless of their weight, fall toward the earth with the same acceleration.

As an old man, Newton himself recounted a version of his apple encounter to younger colleagues. There's no mention of it in materials from around when it would've happened (while he was working on gravity), but he did spend an extended time during that period at his family's farm in Lincolnshire, when the universities were closed due to an outbreak of plague. To the extent that there's truth to the story, however, the most popular telling misidentifies the nature of Newton's insight. Newton's epiphany was not about the existence of gravity but its scope: he realized that the force pulling an apple to the ground is the same force that holds the moon in orbit about the earth, and the earth in orbit around the sun. In the *Philosophiae Naturalis Principia Mathematica*, Newton proposed a universal law of gravitation, giving mathematical form to the attractive force between any two objects in the universe having mass. This form, combined with his laws of motion, allowed physicists to explain the elliptical orbits of the planets in the solar system, the constant acceleration of objects falling near the earth, and a host of other phenomena. It established a template for physics as a mathematical science, one that is followed right up to the modern day.

The crucial feature of Newton's law of gravity is that the force between masses depends on the inverse of the distance between them squared—that is, if you halve the distance between two objects, you get four times the force. Objects that are closer together experience a stronger pull, which explains why the inner planets of the solar system orbit more rapidly. It also means that a diffuse collection of objects will tend to be drawn together, and as they grow closer, they are compressed ever more tightly by the increasing force of gravity.

This increasing force is critical for the continued existence of the sun, and it's the ultimate source of its light. The sun is not a solid object, but rather a vast collection of hot gas, held together only by the mutual gravitational attraction of all the individual atoms making it up. While it may top our list in terms of everyday impact, gravity is the weakest of the fundamental interactions by a mind-boggling amount—the gravitational force between a proton and an electron is a mere 0.00000000000000000000000000000000000001 times the electromagnetic force that holds them together in an atom. The enormous quantities of matter present in the sun, however—some 2,000,000,000,000,000,000,000,000,000,000 kg—build up a gigantic collective gravitational force, pulling everything nearby inward.

A star like the sun begins life as a small region of slightly higher density in a cloud of interstellar gas (mostly hydrogen) and dust. The extra mass in that region pulls in more gas, increasing its mass, and thus increasing the gravitational attraction to pull in more gas still. And, as new gas falls in toward the growing star, it begins to heat up.

At the microscopic scale, a single atom drawn toward a protostar speeds up as it falls inward, just like a rock dropping toward the surface of the earth. You could, in theory, describe the behavior of the gas in terms of the speed and direction of each of the individual atoms, but that's ridiculously impractical even for objects vastly smaller than a sun-sized ball of gas—not just because of the number of atoms, but because the atoms interact with each other. A non-interacting atom would be drawn in toward the center of the gas cloud, speeding up as it went, then would pass out through the other side, slow down, stop, and turn around to repeat the process. Real atoms, though, don't follow such smooth paths: they hit other atoms along the way. After a collision, the colliding atoms are redirected, and some of the energy gained by the falling atom as it accelerated due to gravity is transferred to the atom it hit.

For a large collection of interacting atoms, then, it makes much more sense to describe the cloud in terms of the collective property known as temperature. Temperature is a measure of the average kinetic energy of a material as a result of the random motion of the components making it up—for a gas, this is mostly a function of the speed of the atoms zipping around.* An individual atom is pulled inward and accelerates, gaining energy from gravity and increasing the total energy of the gas. When it collides with other atoms, that energy is redistributed, raising the temperature. The total energy doesn't increase, but rather than having a single fast-moving atom passing through a bunch of slower ones, after many collisions, the average speed of every atom in the sample increases by a tiny amount.

The increasing speed of the atoms in the cloud of gas tends to push it outward, as a faster-moving atom can travel a greater distance from the center before gravity turns it around and brings it back in. The redistribution of energy from new atoms, though, means that this

---

*   To give a sense of the scale, a hydrogen atom in a room-temperature gas is moving at around 600 m/s (about twice the speed of sound), while one near the surface of the sun is moving at about 3,000 m/s.

increase isn't enough to stop the collapse, and as new atoms are pulled in, the mass of the protostar increases, increasing the gravitational force. This, in turn, draws in more gas, bringing in more energy and more mass, and so on. The cloud continues to increase in both temperature and mass, becoming denser and denser, and hotter and hotter.

Left unchecked, the growing force of gravity would crush everything down to an infinitesimal point, forming not a star but a black hole. While these are fascinating objects, warping space and time and presenting a major challenge to our most fundamental theories of physics, the environment near a black hole is not a hospitable place to have a weekday morning breakfast.

Happily, the other fundamental interactions have their own roles to play, halting the star's collapse and allowing the formation of the sun we know and love. The next to kick in is the second most familiar: electromagnetism.

ELECTROMAGNETISM

We regularly encounter the electromagnetic interaction in everyday life, whether in the form of static electricity crackling in a load of socks fresh from the dryer, or that of magnets holding grade-school artwork to the refrigerator. Unlike gravity, which is always attractive, electromagnetism can be either attractive or repulsive—electric charges come in both positive and negative varieties, and magnets have both north and south poles. Opposite charges or poles attract each other, while like poles or charges repel. The electromagnetic interaction is even more ubiquitous than static charges and magnets, though—in fact, it's responsible for our ability to see, well, anything.

In the early 1800s, electromagnetism was a hot topic in physics, with many phenomena involving electric currents and magnets being studied for the first time. Among those studying electromagnetism was British physicist Michael Faraday. He is responsible for a number of technical advances that play key roles in an everyday morning, including his work on liquefying gases, which finds application in refrigeration, and the development of the "Faraday cage" that (among many other things) helps contain the electromagnetic fields used to cook your food to the inside of your microwave oven. Unquestionably, his most important discovery was that not only can electric currents affect nearby magnets, but moving magnets and changing magnetic fields can create current—which is the basis of the vast majority of commercial electricity generation powering the conveniences of modern life. He was one of the first to understand the behavior of charges and magnets in terms of electric and magnetic fields filling empty space and telling distant particles how to move.

Faraday is a seminal figure in physics, one of three people whose likenesses Einstein displayed in his office (the other two were Newton and James Clerk Maxwell). Alas, he came from a poor background, and while he was a great experimenter with deep physical insight, he lacked the formal mathematical training needed to translate this insight into a form that would convince the physicists of his day to take the "field" concept seriously. It fell to James Clerk Maxwell, from a well-off Scottish family, to put electric and magnetic fields on a firm mathematical foundation. In the 1860s, Maxwell showed that all known electric and magnetic phenomena could be explained by a simple set of mathematical relationships—in modern notation, there are four "Maxwell's equations," compact enough to fit on a T-shirt or coffee mug. Faraday's electric and magnetic fields are real things, connected to each other in deep ways—a changing electric field will create a magnetic field, and vice versa.

Maxwell's equations encompass all known electric and magnetic phenomena, and also predicted a new, unified one: electromagnetic waves. If an oscillating electric field is combined in the right way with an oscillating magnetic field, the two will support one another as they travel through space, the changing electric field causing a change in the magnetic field, which causes a change in the electric field, and so on. These electromagnetic waves travel at the speed of light, and light was already known to behave like a wave;* Maxwell's equations were quickly embraced as an explanation for the nature of light—namely, that it is

---

* We'll discuss the experiments that proved the wave nature of light in Chapter 3.

fundamentally an electromagnetic phenomenon. Electromagnetism provides the basis for understanding how light and matter interact, and as we'll see in chapters to come, probing the nature of the interactions between material objects and electromagnetic waves sets the stage for many of the discoveries that established quantum mechanics.

Electromagnetic forces are also largely responsible for the familiar structure of the objects we encounter each day. Ordinary matter is made up of atoms, which are themselves made up of smaller particles distinguished by their electric charge: positively charged protons, negatively charged electrons, and electrically neutral neutrons. An atom consists of a positively charged nucleus containing protons and neutrons, surrounded by a cloud of electrons drawn in by the electromagnetic attraction of the nucleus.

As mentioned earlier, the electromagnetic interaction is vastly stronger than gravity—a fact nicely demonstrated by the party trick of rubbing a latex balloon on your hair and then sticking it to the ceiling. In the rubbing process, a tiny fraction of a percent of the atoms in the balloon steal an electron from atoms in your hair, giving the balloon a small negative charge.* The attraction between this tiny charge and the atoms in the ceiling is strong enough to hold the balloon in place, resisting the gravitational pull of the entire Earth, with a billion billion billion times the mass of the balloon.

The strength of electromagnetism is an indispensable factor in producing the sun. Electromagnetic interactions are responsible for the collisions between atoms that convert the energy gained from gravity into heat. As the temperature of the gas falling into a growing star increases, it eventually becomes hot enough—around 100,000 kelvin or almost 180,000 degrees Fahrenheit†—to separate the electrons in hydrogen atoms from the protons in the nucleus, producing a gas of electrically

---

* Your hair is left with a corresponding positive charge, which is why this trick will make fine hair stand up: the now-positively charged hairs repel each other and spread out as much as they're able to.

† One kelvin is equal to one degree Celsius, but the kelvin scale has no negative numbers and instead starts at absolute zero (the temperature at which molecular activity is at a minimum). Water freezes at 0°C, which is around 273K.

charged particles: a plasma. Gravity continues to compress the plasma, but the mutual repulsion between the positively charged protons holds them apart, resisting gravity's pull. As the forming star continues to draw in more gas, the temperature increases to higher and higher levels.

Despite the enormous disparity between electromagnetism and gravity, though, the plasma can't escape gravity entirely because the electrons that were part of the gas cloud are still around. They're moving too fast to be captured by protons to make atoms, but they keep the star as a whole electrically neutral. If protons alone were present, the mutual repulsion of such an enormous collection of positive charges would blow the whole star apart in an instant. Thanks to the neutralizing background of electrons, though, any individual proton feels only the force of its few nearest neighbors, while the gravitational pull compressing the star comes from the mass of every single particle. As more gas is added, the gravitational force gets stronger and stronger, eventually overwhelming the electromagnetic force.

Electromagnetic interactions can slow the compression of a hot plasma collapsing under gravity, but electromagnetism alone can't stop the collapse and produce a stable star. To create the stable sun as we know it requires an enormous release of energy leading to even higher temperatures, which brings us to the next player in our story: the strong nuclear interaction.

THE STRONG NUCLEAR INTERACTION

The third fundamental interaction is not one that we're directly aware of in everyday life, as it is an extremely short-range force, acting over a distance comparable to the size of an atomic nucleus, about 0.000000000001 mm, or around one ten-billionth the thickness of a human hair. We'd certainly notice its absence, however, as it's responsible for around 99 percent of the mass of everything we deal with.

Understanding the strong nuclear interaction requires us to recognize that two of the particles that make up ordinary matter, protons and neutrons, are in fact pieced together from "quarks," particles with an electric charge equal to a fraction of that of an electron.* A proton is made of two "up" quarks (each with a positive charge two-thirds of the electron charge) and one "down" quark (negative one-third of the electron charge),† while a neutron consists of one up and two down quarks. These quarks are held together by the strong nuclear interaction, similar to the way electromagnetic forces hold electrons in atoms. And just as "electric charge" is the property associated with electromagnetism, the strong force is associated with a property called "color," which takes on three values: red, green, and blue. A three-quark particle like a proton will have one quark of each color, making it "colorless" in the same way that an atom containing equal numbers of protons and electrons is electrically neutral.

The composite nature of protons and neutrons, and the quark-to-quark nature of the strong interaction, helps explain one of the puzzling features of matter, namely how the nucleus of a complex atom holds itself together. Carbon atoms, for example, have six protons in their nucleus, each with a positive charge. As we know from electromagnetism, these positive charges repel each other, producing an enormous force that tries to blow the nucleus apart. So, as many a kid in school learning about atoms has asked, why doesn't the nucleus fall apart?

The answer is the strong nuclear interaction, which, as its name suggests, acts within the nucleus, and is very strong. In fact, it is just over 100 times stronger than electromagnetism, more than powerful enough to hold protons together within an atom. Since the interaction is between individual quarks, though, it only comes into play when the particles are close enough together to "see" that they're made up of quarks. In the same way that two neutral atoms will not interact when they're widely separated but can feel a force pulling them into

---

*   According to our best current understanding, electrons are truly fundamental, and not made up of other, smaller particles.

†   The names "up" and "down" are arbitrary labels, and reflect the tendency of physicists to give things very prosaic names.

a molecule when they get close, colorless protons separated by more than a few times their own radius do not interact with each other via the strong nuclear interaction. The result is similar to the screening of protons by electrons that lets gravity prevent the plasma in a star from blowing apart, as mentioned earlier: the presence of other colors screens out the strong interaction between individual quarks, leaving only the electromagnetic repulsion.

Close up, however, the individual quarks in neighboring particles are drawn to each other, and this is what holds protons (and neutrons) together inside the nucleus. This is also where the strong interaction comes into play within the sun. At ordinary temperatures, electromagnetism keeps protons too far apart for the strong interaction to kick in, but as the plasma inside a forming star gets hotter and hotter, and protons move faster and faster,* they begin to approach each other more and more closely. At the temperature and density of matter found in the core of a star-to-be, a tiny fraction of these protons will get close enough for the strong force to take over and stick them together. This process converts hydrogen (the simplest atom, with a nucleus containing a single proton) into helium (a nucleus with two protons and two neutrons), and along the way releases an enormous amount of energy.

Where does this energy come from? The short answer is "the world's most famous equation, $E = mc^2$." That is, some of the mass of the initial hydrogen is converted into energy: the energy release of the sun involves converting four million metric tons of mass into energy every second. But that answer can be kind of confusing, since the total number of particles doesn't change—four hydrogen nuclei contain twelve up and down quarks, as does a helium nucleus—so it's not obvious where the missing mass came from. Explaining this requires a deeper look inside the proton and the nature of the strong interaction.

---

* The electrons also move faster, but they were already moving at such high speeds that the increase doesn't make much difference. Their only role in the plasma inside a star is to provide a diffuse background of negative charge, keeping the whole star electrically neutral.

Particle physicists have been aware of the existence of quarks since the 1960s, and the properties of the up and down quarks are well-known. If you search for "quark" using Google, you'll get all manner of information about these particles, including the masses of up and down quarks—2.3 and 4.8 in the units physicists use to measure such things.* This is a little surprising, though, as the mass of a proton in those same units is 938, about 100 times greater than the mass of the particles that make it up.

So, where does the mass of a proton come from? The answer, again, is $E = mc^2$: The quarks inside the proton are bound together by the strong nuclear interaction, and that interaction involves an enormous amount of energy. To observers outside the proton, this interaction energy manifests as mass. Something like 99 percent of the mass of a proton, then, is not in the form of material particles, but is energy from the strong interaction holding the proton together.

The same basic process takes place inside an atom, between the protons and neutrons bound together by the strong force. The mass we measure for the nucleus of an atom is not just the sum of the masses of the protons and neutrons making it up, but it also includes a contribution from the energy of the strong interaction that binds them together.

Exactly *how much* mass the strong interaction contributes, though, depends on the details of the specific atom and how it's put together. For very light atoms like hydrogen and helium, it turns out to be slightly more efficient to have a bigger nucleus—the amount of strong-interaction energy needed to keep two protons and two neutrons together is slightly less than that required for four individual

---

* The units are based on the energy content through $E = mc^2$: an up quark's mass of 2.3 MeV/C^2 indicates that converting an up quark into energy would release 2.3 million electron volts worth of energy (the usual way this happens is when an up quark annihilates with its antimatter equivalent, releasing two photons each with 2.3 MeV of energy). Or, to flip it around, it requires 2.3 MeV of collision energy to create an up quark in a particle accelerator (or, more practically, 4.6 MeV to create an up quark and an up antiquark as a pair).

protons. When four protons undergo nuclear fusion to make helium,* then, they no longer need some of the energy they initially had, and that energy gets released as heat. The energy released per reaction is very small—a baseball pitched with this same energy would take about a month to reach home plate—but there are vast amounts of hydrogen fusing inside the sun, a staggering 10^{38} (1 followed by 38 zeroes) of these reactions happening every second (give or take).

To recap: As a star like the sun is forming, gravity and electromagnetism begin the process of heating the gas as it falls toward the center. When the temperature gets high enough for a few hydrogen atoms to begin fusing into helium, the energy they release rapidly increases the temperature, which in turn increases the rate of fusion. Eventually, a balance is reached between the inward pull of gravity and the outward pressure produced by this heating, and the star remains stable for as long as there is hydrogen in the core to "burn."

The several-billion-year life of a star, then, is driven by gravity, electromagnetism, and the strong force. Gravity draws gas together, electromagnetism resists the collapse and increases the temperature, and when that temperature is high enough that electromagnetism no longer keeps protons well apart, the strong nuclear force releases vast amounts of energy as hydrogen fuses into helium. The competition between the three produces a stable star, generating the light and heat that sustains life on our planet.

It might seem like we've told the entire story with only three of the four fundamental interactions, sadly neglecting the weak nuclear interaction (which was already saddled with the worst name of the lot). But in fact it, too, has a part to play in powering the sun—a contribution more subtle than that of the others, but no less essential.

---

* If you look in detail at the hydrogen fusion process, it's rather complicated, with multiple possible intermediate paths involving interactions with additional particles and the temporary formation of unstable elements. In the big picture, though, what matters is just the energy difference between the start state (four free protons) and the end state (a helium nucleus).

THE WEAK NUCLEAR INTERACTION

The weak nuclear interaction occupies an unusual position in the Standard Model, being arguably the least obvious fundamental interaction, while also being one of the best understood. The mathematical theory of the weak interaction and its close relationship with electromagnetism was developed through the 1960s and early 1970s, and the experimental confirmation of that theory's predictions, culminating in the discovery of the "Higgs boson" in 2012, ranks among the greatest triumphs of the Standard Model. The strong nuclear interaction, meanwhile, continues to pose problems for theorists computing properties of matter, while gravity is famously mathematically incompatible with the other three.*

At the same time, however, it's very difficult to point to exactly what the weak nuclear interaction *does*. What makes the weak interaction especially tricky to explain to nonphysicists is that, unlike the other interactions, it doesn't manifest as a tangible force in the usual sense. The pull of gravity is a central element of our everyday experience, and electromagnetic forces between charges and magnets are something you can feel. And while the strong interaction operates at an extremely remote scale, it's still easy to understand as a force holding the nucleus together against electromagnetic repulsion.

The weak interaction, on the other hand, isn't used to stick anything together, or to push anything apart. This is why most physicists have dropped the pleasingly alliterative term "fundamental forces"

---

* Our best theory of gravity is general relativity, which describes gravity's effects in terms of a curvature of space and time, which are smooth and continuous. The other three forces are described by quantum theories that involve discrete particles and sudden fluctuations. The mathematical techniques used for one do not easily translate to the other, and the problem of finding a way to combine them to make a quantum theory of gravity has bedeviled theoretical physics for decades. Happily for us, the situations where you need both quantum physics and general relativity are very rare—near the center of a black hole, or in the very early universe—and not the sort of thing you'll see in the course of a typical morning.

in favor of "fundamental interactions." Instead of pushing or pulling on particles, the weak nuclear interaction's important function is to cause particle transformations: more specifically, it turns particles from the quark family into particles from the lepton family. This lets a down quark (which has a negative charge) transform into an up quark (which has a positive charge) by emitting an electron and a third particle known as a neutrino—or an up quark transform into a down quark by absorbing an electron and emitting an antineutrino. These transformations enable neutrons to turn into protons, and vice versa.

The process taking place in the sun involves the latter, and is the inverse of the better-known phenomenon of "beta decay," in which a neutron in the nucleus of an atom spits out an electron and changes into a proton. Beta decay has been known of since the early days of research into radioactivity, but explaining it posed a vexing challenge in the early days of quantum theory, leading to one of the more colorful anecdotes of twentieth-century physics.

The problem with beta decay is that the electrons spit out by decaying nuclei emerge with a wide range of energies (up to some maximum value). This shouldn't be possible for a reaction involving only two particles—the laws of conservation of energy and conservation of momentum should dictate only a single possible energy for the departing electron (as is the case for the process of "alpha decay," in which a heavy nucleus decays by spitting out a helium nucleus: two protons and two neutrons stuck together). Explaining the range of energies seen in beta decay stymied physicists for a long time and led some to propose drastic measures—like abandoning the idea of conservation of energy as a fundamental principle of physics.

The solution was found by the young Austrian physicist Wolfgang Pauli, who in 1930 suggested (in a letter sent to a conference he was skipping to attend a ball in Zurich) that beta decay *didn't* involve two particles, but *three*—the neutron-turned-proton, the electron, and a third, undetected particle with a very tiny mass. The new particle, quickly dubbed the "neutrino" (loosely "little neutral one" in Italian), carries away some energy, with the precise amount depending on the exact momentum of the electron and neutrino when they leave the nucleus.

Introducing the neutrino initially didn't seem much less desperate than ditching conservation of energy—Pauli himself wrote to a friend, "I have done something terrible. I have postulated a particle that cannot be detected. That is something a theorist should never do." Within a few years, though, the great Italian physicist Enrico Fermi developed Pauli's rough suggestion into a complete and remarkably successful mathematical theory of beta decay, and the idea was quickly adopted. Pauli's original neutrino turns out to be one of three (the original electron neutrino, plus "muon" and "tau" varieties), and despite his initial lament, neutrinos can, in fact, be detected, and were experimentally confirmed by Clyde Cowan and Frederick Reines in 1956.[*]

What does all this have to do with the sun? The answer is subtle, but hinted at a few times in the earlier discussion of fusion. The sun is powered by fusing hydrogen nuclei, which are single protons, into helium nuclei consisting of two protons and two neutrons stuck together. Somewhere in this process, two protons need to turn into neutrons, which is possible thanks to the weak nuclear reaction and the process of "inverse beta decay" mentioned above, in which a proton turns into a neutron, emitting a neutrino in the process.[†] As a result, the sun produces incredible numbers of neutrinos, which have been detected on Earth, and measurements of these solar neutrinos provide information both about nuclear reactions in the core of the sun, and about the properties of neutrinos themselves.

The conversion of protons to neutrons inside stars is essential for the existence of the enormous range of elements we interact with on

---

[*] Reines shared the 1995 Nobel Prize in Physics for this work (Cowan had died in 1974, and the Nobel is not awarded posthumously), and two other Nobels have been given for work with neutrino detectors: to Raymond Davis Jr. and Masatoshi Koshiba in 2002, and to Takaaki Kajita and Arthur B. McDonald in 2015.

[†] In the process, the proton must also either emit a positron (the antimatter equivalent of the electron) or absorb one of the huge number of electrons left over from the original gas present in the sun. Any positrons emitted will quickly annihilate with one of the aforementioned electrons, so the end result is the same from the outside: one proton and one electron are gone, leaving one neutron and one neutrino in their place.

a daily basis—the oxygen in the air we breathe and water we drink, the carbon in the food we eat, the silicon in the ground beneath us. When a very heavy star burns through most of the hydrogen in its core, it begins to fuse helium into even heavier elements; when the helium runs low, these extremely heavy stars begin to burn carbon, and on up through the periodic table. At each step of the process, though, the strong-interaction energy released by fusion decreases,* until silicon is fused into iron. The fusion of iron does not release any energy, cutting off the flow of heat that's propping up the core of the star. At that point, the outer layers of the star come crashing inward, and bounce off the core to produce a supernova explosion, releasing enough energy that the exploding star often temporarily outshines the rest of its home galaxy.

In a supernova, much of the mass of the star is blasted outward in an expanding cloud of gas, carrying with it the heavier elements produced in the core during the later stages of fusion. These gas clouds expand and cool and interact with other gas in the neighborhood, forming the raw material for the next generation of stars—and also rocky, Earthlike planets, which are largely made up of the heavy elements created in the core of the dying star.

The enormous variety of substances we see on Earth—rocks and minerals, breathable air, plants and animals—are all built from the ashes of dead stars, created through all four fundamental interactions. Starting with simple clouds of hydrogen formed shortly after the Big Bang, gravity pulls gas together, electromagnetism resists the collapse and

---

* The decreasing energy return can be understood in terms of the strong-force energy that manifests as mass: the energy required to hold twelve quarks together in a helium nucleus is substantially less than that needed for four individual, unattached protons, but as the number of particles increases, the energy savings from adding new ones to the mix decreases. It's a little like the organizational efficiency of grouping people: two people sharing an apartment can live more cheaply together than alone, but adding roommates only saves money up to a point. The hassle of accommodating a sixth roommate can be greater than the cost savings can justify. In a similar way, the energy saved by adding more particles to a large nucleus just isn't that much.

heats the gas, and the strong nuclear interaction releases vast amounts of energy in nuclear fusion. And, finally, the weak nuclear interaction enables the particle transformations that turn hydrogen into heavier and more interesting elements. Take any one of these fundamental interactions away, and our everyday existence would be impossible.

THE REST OF THE STORY

The above is not by any means the complete story of fundamental physics. The four fundamental interactions that power the sun are the only ones we know of, but the Standard Model includes four types of quarks beyond the up and down varieties that make up protons and neutrons, and four additional leptons beyond the electron and electron neutrino. The particles of the Standard Model also have antimatter equivalents—particles with the same mass but the opposite charge—and when a particle encounters its antimatter equivalent, they mutually annihilate, converting their mass into high-energy photons of light. All of these particles have been experimentally confirmed, and their properties studied in great detail.

None of these additional particles stick around for long, though—the longest-lasting is probably the muon, with an average lifetime of around two one-millionths of a second—so their influence on everyday experience is pretty minimal. They're created for a fleeting instant in high-energy collisions between more ordinary particles, whether in earthbound physics experiments or astrophysical events, and they decay rapidly into up and down quarks (usually in the form of protons and neutrons), electrons, and neutrinos. The history of their discovery and the development of the Standard Model is a fascinating story, but one beyond the scope of this book.

For the purposes of exploring the physics of everyday objects, we can largely confine ourselves to just the three most familiar material particles: protons, neutrons, and electrons. These combine to make atoms, which in turn make up everything we interact with in the course of an ordinary day. In terms of fundamental interactions, a typical morning routine relies mostly on electromagnetism, which is responsible for holding atoms and molecules together, and connecting matter to light.

It's worth remembering, though, that deep beneath the surface, even something as seemingly fundamental as the mass of objects can be traced back to the exotic physics of the strong nuclear interaction. And that all four interactions, acting among an assortment of quarks and leptons, are required for the operation of even our most quintessentially everyday companion—the sun.

CHAPTER 2

THE HEATING ELEMENT: PLANCK'S DESPERATE TRICK

*Down in the kitchen, I put water on for tea—checking for **the glow of the heating element** to make sure I haven't groggily put the kettle on the wrong burner again . . .*

The red glow of a hot object is one of the simplest and most universal phenomena in physics. If you get a chunk of material—any material—hot enough, it will start to glow, first red, then yellow, then white. The color of the light emitted depends *only* on the temperature of the object. The material used doesn't matter—a rod of clear glass and one of black iron, heated to the same temperature, glow with exactly the same color. The method of heating doesn't matter, either—whether you're running an electric current through a coil of metal, as in my electric stove, or forging that coil in a fiery furnace, the color of the hot metal will be the same for that particular temperature.

This sort of simple and universal behavior is like catnip for physicists because it suggests that there should be some simple and universal underlying principle at work. In the late 1500s, Galileo Galilei and Simon Stevin demonstrated empirically that objects of different materials and weights all fall at the same rate—Stevin by dropping two lead balls, one ten times heavier than the other, from a church tower.[*] This observation led Isaac Newton to develop his law of universal gravitation in the 1600s, and a few hundred years after that, a different perspective on the same simple, universal behavior inspired Albert Einstein's general theory of relativity, which remains our best theory of gravity. Einstein recalled the key moment in the development of his theory as an afternoon in 1907 at the patent office in Bern, when he was struck by the realization that a person falling off a roof would feel weightless during the fall, an insight that provided the link between acceleration and gravity that is the foundation of general relativity. Einstein referred to this as "the happiest thought of my life." Working out the consequences of that happy thought mathematically took the better part of eight years, but culminated in one of the greatest and most successful theories in modern physics.

The universal behavior of thermal radiation, then, seemed like a similarly promising source of insight, a phenomenon against which to test ideas about the distribution of energy in hot objects and the ways light and matter interact. Unfortunately, the best efforts of physicists in the late 1800s to predict the color of light emitted by hot objects at different temperatures failed spectacularly.

In the end, a full explanation of thermal radiation required a radical break with existing physics. The starting point for the whole of quantum theory, whose implications physicists are still debating more than

---

[*]   This works provided both objects are dense enough for air resistance forces to be negligible—if you were to drop a paper clip and a feather, the paper clip would drop rapidly while the feather would flutter to the ground slowly. The force of gravity acting upon them, though, is the same—in a vacuum, they would reach the ground together, as demonstrated dramatically by Commander Dave Scott during the Apollo 15 mission to the moon.

a century later, is found in the red glow of the heating elements we use to cook breakfast.

In a very real sense, then, all of the bizarre phenomena associated with quantum physics—particle-wave duality, Schrödinger's cat, "spooky action at a distance"—can be traced back to your kitchen.

LIGHT WAVES AND COLORS

As is often the case, the easiest way to explain the need for a radical new theory is to first illustrate the failure of the old one. Before we can understand how the quantum model solved the problem of thermal radiation, we need to see why classical physics *couldn't*. That, of course, requires a bit of background in what classical physics had to say about light, heat, and matter.

The first essential concept underlying the experiments that led to the breakdown of classical physics is the idea of light as a wave. The wave nature of light was known for a half century before Maxwell's equations, thanks in large part to experiments carried out around 1800 by the English polymath Thomas Young. Physicists had been arguing about whether light was best thought of as a stream of particles or a wave through some medium since the days of Newton, but Young convincingly demonstrated light's wave nature with his ingeniously simple "double-slit" experiment.

As the name suggests, the double-slit experiment involves light passing through two narrow openings cut in a card. Young found that shining light through two closely spaced slits to a screen on the other side does not result in the two bright stripes you might expect (for light passing through each individual slit). Instead, what appears on the screen is a series of bright and dark spots.*

---

* If you'd like to see for yourself, you can make two fine slits in a piece of aluminum foil and illuminate them with light from a laser pointer. Another closely related phenomenon is even easier to see: if you put a strand of hair in

These spots arise from a process known as "interference," which occurs whenever waves from two different sources combine. If the two waves reaching a given point arrive "in phase," so that the peaks of one wave coincide with the peaks of the other, the waves combine to form a wave with a higher peak than either had alone. On the other hand, if the waves arrive "out of phase," with one at a peak when the other is in a valley, they cancel out: the peaks of one fill in the valleys of the other, and the end result is no wave at all. This works with any source of waves—it's responsible for the complex patterns of waves seen in wave pools at amusement parks, and the destructive interference of sound waves is the basis for "noise canceling" headphones.

The interference in Young's double-slit experiment comes about because light waves from each slit take different amounts of time to travel to a particular point on the screen. At a point exactly centered between the two slits, both waves travel the same distance, and thus arrive in phase, giving a bright spot. At a point a bit to the left of the center, the waves from the left-hand slit travel a shorter path to the screen than that taken by the waves from the right-hand slit. This extra distance means the waves from the right slit have had a bit more time to oscillate, and if the distance is just right, the peaks of the right-slit waves fill in the valleys of the left-slit waves, making a dark spot. A bit farther out, though, the extra distance allows for an extra full oscillation, putting the right-slit peaks on top of the left-slit peaks, and making another bright spot.

This pattern repeats many times, leading to the array of bright and dark spots. The spacing between bright spots depends in a simple way on the wavelength, providing a convenient way to measure the wavelength of visible light—in modern units, this ranges from about 400 nanometers for violet light to about 700 nanometers for deep red.* Adding more slits makes the bright spots narrower and more distinct, and by the 1820s, Joseph von Fraunhofer was using "diffraction gratings"

---

the beam of a laser pointer, the light waves passing around the different sides of the hair will interfere and make a pattern of multiple spots.

*   One nanometer is 10^{-9} m, or 0.000000001 m.

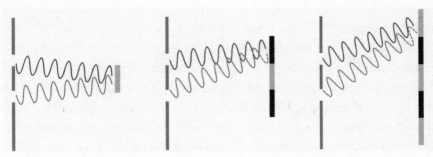

Interference of light waves in a double-slit experiment. Midway between the slits, the waves arrive in phase and combine to create a bright spot. A bit above the center, waves from the bottom slit travel a longer distance, and thus undergo an extra half oscillation (dashed line), so that the peaks from the bottom wave fill in the valleys from the top, producing a dark spot. Some distance farther out, waves from the bottom slit complete a full additional oscillation (dashed line), and the waves are once again in phase, producing another bright spot.

based on the interference of light to make the first reasonably precise measurements of the wavelengths of light emitted by the sun and other stars.

Young's experiment, published in 1807, caused some sensation in physics circles, but many scientists remained reluctant to discard the particle theory of light. When the French physicist Augustin-Jean Fresnel submitted a paper on wave theory to a physics competition, one of the holdouts, Siméon Denis Poisson, pointed out that the wave interference used to explain Young's experiment would predict that there should be a bright spot at the center of the shadow of a round object. This bright spot in a shadow seemed plainly absurd, and Poisson thus rejected the wave model of light.

François Arago, one of the judges for the competition, was intrigued by Poisson's idea and began a careful experimental search for bright spots at the center of shadows. Observing the spot takes exceptional care, but Arago was up to the task, and he definitively demonstrated that light passing around a circular obstacle really can interfere to produce a bright spot at the center of the shadow. This "spot of Arago" or

"Fresnel spot" was the final bit of evidence needed to convince most physicists that light was indeed a wave.

Arago's experiment secured the success of the wave model, but exactly *what* was waving remained a mystery into the 1860s, when Maxwell's equations explained light as an electromagnetic wave. In the closing decades of the 1800s, then, the wave theory was firmly established, and physicists were seeking to explain all interactions between light and matter in terms of electromagnetic waves.

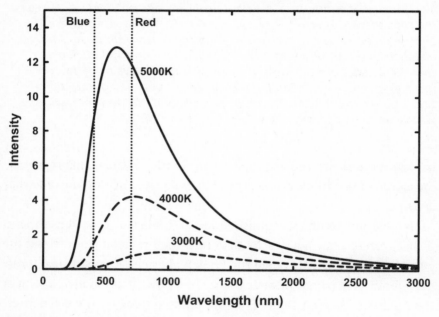

The spectrum of thermal radiation for several different temperatures. The vertical lines indicate the limits of the visible spectrum, showing how the peak moves from the infrared into the visible with increasing temperature.

When studying waves, there are two properties that we can readily measure: the wavelength and the frequency. Wavelength is the distance between peaks in the wave, looking at a snapshot of the whole pattern over some region. Frequency is the time between peaks measured from a single point watching the wave go by. Because light travels at a fixed speed, frequency and wavelength are closely related: the wave moves forward one wavelength for each oscillation. Shorter wavelengths repeat more often over the same period, so they have higher frequencies. Physicists switch back and forth between talking about light in terms of frequency and in terms of wavelength depending on which is most convenient for the particular problem at hand—we'll make this switch a couple of times in the rest of this chapter.

Determining the "color" of light emitted by a hot object is a matter of measuring its spectrum: the intensity of light emitted at each frequency over a wide range. When we measure this spectrum for light at a particular temperature, we find a simple characteristic shape, a distribution with lower amounts of light at the lower frequencies, increasing to a peak, and then dropping off rapidly at the higher-frequency end. The "color" of the light is determined by the position of this peak—the exact frequency at which the emitted intensity is greatest—and depends only on the temperature, in a very simple way. As the temperature increases, the frequency at which the amount of light emitted reaches a maximum gets higher: at room temperature, the peak intensity is in the far infrared region of the spectrum, moving into the red end of the visible spectrum as the temperature increases to "red-hot," and toward the blue end as the temperature increases further. A "white-hot" object has the peak of its spectrum in a region that would correspond to green light,* but it emits significant amounts of light across the entire visible range of the spectrum and thus looks white. If you double the temperature (measured on the kelvin scale, which begins at absolute zero), the peak frequency also doubles.

The spectrum of light from the sun closely resembles this universal spectrum for light from a hot object, corresponding to a temperature of about 5600K, peaked at a frequency of around 600 THz—in fact, this is how we measure the temperature of the sun and other stars. At the other extreme of temperature is the cosmic microwave background,

---

*   Relating the wavelength or frequency of light to the color perceived by humans is a tricky business, particularly when it comes to dealing with light at multiple frequencies. The color addition that kids learn in elementary school is an example of this—a mix of red light (around 650 nm wavelength) and blue light (around 490 nm) will create the same impression in your eyes and brain as violet light (around 405 nm) even though there is no violet light present.

relic radiation left from shortly after the Big Bang that permeates the universe with a spectrum corresponding to that of an object at 2.7K, peaked at around 290 GHz.

HEAT AND ENERGY

Throughout the nineteenth century, in parallel with the developments in theories of electromagnetism and the wave model of light, there were great advances in the physics of thermodynamics. Just as the century opened with debate over two models of light—wave and particle—the early decades of the 1800s also saw debate over two competing models of heat. One school of thought viewed heat as a physical thing unto itself—a "subtle fluid" called "caloric" that flowed from one object to another. The competing model, "kinetic theory," envisioned heat as arising from the random motion of the microscopic components making up macroscopic matter.

Over a period of several decades, experiments by Benjamin Thompson (also known as Count Rumford) and James Joule demonstrated a connection between mechanical work and the generation of heat that was difficult to reconcile with the caloric theory. Thompson showed that the friction involved in boring out a cannon could provide a seemingly inexhaustible source of heat, which should not have been possible if "caloric" were a real fluid. Joule strengthened this relationship by determining a precise value for the "mechanical equivalent of heat"— that is, how much work was needed to raise the temperature of a fixed amount of water one degree by stirring it.

On the more theoretical side, work by Rudolf Clausius and James Clerk Maxwell* established the mathematics linking the flow of heat between objects to the kinetic energy of the atoms and molecules making them up. The Austrian physicist Ludwig Boltzmann built on

---

* Yes, the same Maxwell who worked on electromagnetism. Physics in Europe in the 1800s was a smallish community, and Maxwell was a really smart guy.

Maxwell's work, developing much of the statistical model of heat energy that we use today.

Individual atoms and molecules in a gas or solid rattle around at different velocities, but given a large enough number of them, we can use statistical methods to precisely predict the probability of finding atoms with a certain kinetic energy in a substance at a specified temperature. (The resulting formula is known as the "Maxwell-Boltzmann distribution" in honor of their pioneering work.) A crucial piece of this kinetic model is the notion of "equipartition," introduced by Maxwell and refined by Boltzmann, which holds that energy is distributed equally among all types of motion available to a particle. A gas of single atoms has all its kinetic energy contained in the linear motion of its atoms, while a gas of simple molecules will have its energy split equally between linear motion of the molecules as a whole, vibration of the atoms within the molecules, and rotation of each molecule about its center of mass. Kinetic theory and this statistical approach successfully explained the thermal properties of many materials,* and by the end of the 1800s, caloric theory had fallen by the wayside.

Since the emission of light requires heat energy, and light plays a significant role in transmitting heat—this is why cooks cover some dishes with foil, to block light and reduce burning—physicists naturally began to investigate the connection between electromagnetic waves and thermal energy. This project required empirical data, so in the late 1800s, spectroscopists in Germany conducted experiments to measure the spectrum of light emitted by hot objects over a wide range of temperatures and wavelengths. The experimental results were of high quality, but an explanation of those results in terms of the kinetic model of thermal physics remained elusive.

In the 1890s, two competing models, by Wilhelm Wien in Germany and Lord Rayleigh in Britain, made empirical predictions of the amount

---

*   At high temperatures, anyway; at very low temperatures, and for some very hard materials, the Maxwell-Boltzmann kinetic theory fails. These anomalies were another hint of the need for new physics and would play a role in the rise of quantum mechanics in the early 1900s.

of light emitted at a given wavelength for a given temperature—formulae based on general principles and experimental data from one range of wavelengths that they hoped to extend to other ranges. Wien's predictions matched the data at high frequencies but failed at lower ones, while Rayleigh's worked only at low frequencies. In 1900, Max Planck found a mathematical function that combined the two and at last lined up with the observed data. Planck derived this function after a party he hosted where spectroscopist Heinrich Rubens told him about Rayleigh's predictions and the latest experimental results. When the guests left, Planck retreated to his study, and some time later emerged with the correct formula, which he sent to Rubens on a postcard the same evening. But while Planck's formula was a great empirical success, nobody could explain *why* it worked, at least not using what, at the time, were the accepted fundamental principles of physics.

THE ULTRAVIOLET CATASTROPHE

So, what should a model based on those principles look like? The general approach is most clearly illustrated by the method attempted by British physicists Lord Rayleigh and James Jeans (which actually slightly post-dates Planck's successful quantum model). The Rayleigh-Jeans model fails, but in a way that makes the origin of the failure clear, and the eventual solution can be explained using the same basic language.

The idea behind the Rayleigh-Jeans approach to the problem of thermal radiation is very simple, and relies on the notion of equipartition used by Maxwell and Boltzmann in describing the thermal properties of gases: you simply take the energy available from heat and divide it evenly among the possible frequencies of light. "Divide it evenly" demands a countable set of possible frequencies, though, which means physicists would need a simplified theoretical model to break down the continuous spectrum of light.

The trick to making the frequencies countable follows directly from the universality of the radiation observed: remember, the spectrum of light from a hot object doesn't depend on any of the material properties of that object. The theoretical model would need to reflect this, which led physicists to consider the light emitted by an idealized "black body," an object that absorbs any and all light that falls on it, reflecting nothing.* This doesn't mean that the object is dark, emitting no light—if that were the case, it would rapidly heat up and disintegrate—only that, as with the glow of a heating element, the light it emits does not depend in any way on the light it absorbs.

It turns out that there's a nice, practical way to make such a black body in the lab: a box with a small hole in it. As long as the hole is small compared to the size of the box, any light entering will be extremely unlikely to come right back out, but will instead have to bounce around many times before it manages to escape (if it isn't absorbed first). This approximates the essential "blackness" of the black body: light falling on it is absorbed and not reflected, regardless of frequency. The physicists making measurements of thermal radiation† used exactly this technique to make the sources for their experiments.

The box-with-a-small-hole model is also a great boon for theoretical physicists because the waves inside the box will be restricted to a limited set of frequencies. Waves that fit nicely within the boundaries of the box endure, while waves at the "wrong" frequencies will interfere with each other and get wiped out. Whatever light leaks out of the hole, then, will reflect the limited set of frequencies that exist inside, and have nothing to do with whatever's going on outside the box.‡

Once physicists hit upon the trick for determining a limited set of allowed frequencies, the hope was that when they tallied up the allowed frequencies inside the box, and divided the available energy among them, the resulting spectrum would resemble that observed in experiments and described by Planck's formula. Unfortunately,

---

* In the immortal words of Nigel Tufnel in *This Is Spinal Tap*, "How much more black could this be? And the answer is: None. None more black."

† Notably the German experimentalists Otto Lummer and Ferdinand Kurlbaum.

‡ This might seem like it's reintroducing properties specific to a particular "box," but as long as the box is very large compared to the wavelength of the waves inside, there are well-established mathematical techniques for smoothing this out to get an answer that doesn't involve the size of the specific box.

this simple and straightforward approach failed spectacularly. We can see the problem just by going through the process of counting up the allowed frequencies.

The allowed frequencies inside the box are called "standing-wave modes," and these are determined by the size of the box and the constraint that none of the waves are allowed to leave (as long as the hole in the box is small enough, the fraction of light that escapes is so tiny it can safely be ignored). For the sake of illustrating the origin and characteristics of these standing-wave modes, we can simplify things still further, imagining a "box" that has just one dimension: waves can travel only left and right, no other directions. This has a simple and familiar everyday analogue: the string of a musical instrument.

A guitar player makes sound by plucking a string, displacing a small part of the string and creating a disturbance that travels outward in the form of waves shaking the string up and down. The two ends of the string are fixed, so when a wave traveling up the neck reaches the player's finger pressing the string against the fret, it bounces back, reversing direction to travel back down the neck again. It doesn't take long before waves traveling in opposite directions find themselves occupying the same stretch of string, at which point they interfere with each other like the light from the two slits in Young's famous experiment.

When you add together all these waves bouncing back and forth, you find that for most wavelengths, the end result is complete destructive interference. For every wave trying to make the string rise to a peak, there's another trying to push it down to a valley, and they cancel each other out. For a very particular set of wavelengths, though, you get constructive interference: all the various reflected waves rise to a peak at exactly the same place. These wavelengths give rise to stable patterns of waves along the string, where some parts of the string move quite a bit, while others remain fixed in place.

The simplest of these patterns is a "fundamental mode" with a single oscillating lump between the fixed ends. We typically draw this as an upward-going bump, but really it varies in time: the bit of string in the middle is pulled upward, then it drops back to the flat position, then it moves down to a negative peak, then back to zero, then back to the upward peak, and so on. The time required to complete an oscillation is determined by the frequency associated with the wavelength of the mode in question.

The wavelength of a wave is defined as the distance to go up to a peak, then down to a valley, and back to the start. A single up-and-back-to-zero motion is half a wave, so the wavelength associated with the fundamental mode is twice the length of the string. The next simplest pattern fits a full wave between the fixed ends, going up (or down) and then back down (or up), with a fixed "node" in the center where the string does not move; the wavelength of this second "harmonic" is exactly equal to the length of the string. The next harmonic has one and a half waves (three oscillating lumps and two nodes) for a wavelength of two-thirds the length of the string; the next has two waves with a wavelength of half the length of the string, and so on.

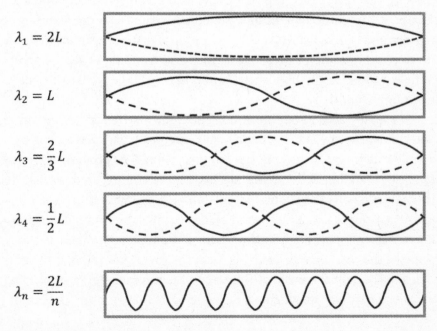

$\lambda_1 = 2L$

$\lambda_2 = L$

$\lambda_3 = \dfrac{2}{3}L$

$\lambda_4 = \dfrac{1}{2}L$

$\lambda_n = \dfrac{2L}{n}$

Some of the standing-wave modes in a one-dimensional "box" of length L with the wavelength λ for each mode.

If we look closely at these allowed modes, we find a simple pattern: in each of the allowed standing-wave modes, a whole number of half wavelengths fit across the length of the string. There are a discrete set of these allowed modes, and we can assign each of them a number—the number of oscillating lumps in the pattern.

The sound that we hear from a guitar makes a nice analogy with the spectrum we see from a black body in this model. The initial plucking of the string will excite waves at a huge number of different frequencies, like the light that enters the box for our black body. After a very short time, though, destructive interference between the many reflections off the ends of the string or the walls of the box wipes out most of these wavelengths, leaving only those that correspond to standing-wave modes.

In the case of the guitar string, most of the energy of the wave ends up in the fundamental mode, which as the name suggests is the primary determinant of the sound that we hear. The higher frequency harmonics get a smaller share of the energy but are still present, and they are responsible for the rich sound of a real instrument compared to, say, a computer generating a single pure tone. The many different tunings and effects used by guitarists produce distinctly different tones by amplifying some of these harmonics and damping others to give a different mix that distinguishes the sound of, say, Jerry Garcia's guitar from Jimi Hendrix's.

For light waves in our black-body box, the distribution of energy is determined not by the aesthetic tastes of a particular player, but by a simple rule from thermal physics: equipartition. The process of identifying the standing-wave modes is a little more complicated for light in three dimensions than sound in one dimension, but leads to the same result: a discrete set of numbered modes that can be counted. Once we know these modes, equipartition tells us to allot each mode an equal share of the total energy available from the thermal motion of the particles making up the walls of the box (which, remember, are standing in for the particles making up a hot object).*

---

* Admittedly, there is an infinite number of these modes, but dealing with these kinds of infinities is exactly the reason physicists invented calculus.

The problem is that as the wavelengths get shorter, the wavelengths of allowed modes get closer and closer together. If we tally up the number of modes within some given range of wavelengths, we find that it increases without limit at short wavelengths (which, remember, correspond to high frequency). If we imagine a string half a meter long, with a fundamental wavelength of one meter, there are two allowed modes with wavelengths in the five millimeters between 0.1 m and 0.095 m—that is, two wavelengths that can fit an integer number of their half wavelengths across the string. In the five-millimeter wavelength range between 0.02 m and 0.015 m, there are thirty-four modes. Between 0.01 m and 0.005 m, there are over two hundred modes.

In terms of a spectrum, this model doesn't reproduce the nice, simple peak at an intermediate wavelength found in experiments. On the contrary, it says that any object, regardless of temperature, ought to spray out an infinite amount of short-wavelength (high-frequency) radiation. This is *not* what you want in a toaster.

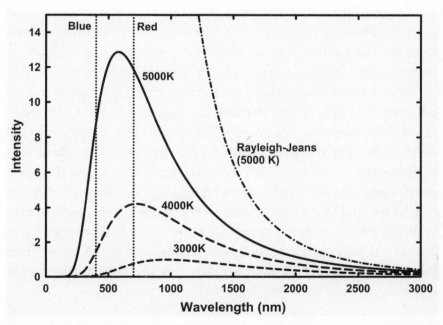

The spectrum of thermal radiation at different temperatures, plus the prediction of the Rayleigh-Jeans model, i.e., the "ultraviolet catastrophe."

This failure of the straightforward mode-counting approach was so bad that it picked up the name "ultraviolet catastrophe."* Explaining the peak seen in the real black-body spectrum, and successfully described by Planck's formula from 1900, required a fundamental shift in our understanding about the way energy is distributed.

THE QUANTUM HYPOTHESIS

Fittingly, it was the same Max Planck who'd found the mathematical function accurately describing the shape of the spectrum of emitted light who also eventually found a way to explain the origin of that spectrum. In the terms of the model described above, Planck associated each of the standing-wave light modes with an "oscillator" inside the material, with each oscillator emitting only a single frequency of light. He then assigned each of these oscillators a characteristic energy, equal to the frequency of that oscillator multiplied by a small constant. Then he required the amount of light emitted by a given oscillator to be an integer multiple of its characteristic energy, which he called a "quantum" after the Latin word for "how much"—so an oscillator could have one quantum of energy, or two, or three, but never half a quantum, or π quanta.

This "quantum hypothesis" does the necessary trick of cutting off the amount of light at high frequencies—exactly where the ultraviolet catastrophe happens. When we allocate each "oscillator" an equal share of the heat energy available, for low-frequency oscillators, that share amounts to many times its characteristic energy, and thus each emits many quanta of light. As the frequency increases, the amount of light emitted by each individual oscillator goes down, because each oscillator's share of the heat energy amounts to a smaller multiple of its characteristic energy. And when the frequency gets high enough that the characteristic energy is bigger than that oscillator's share of the heat energy, it can't emit any light at all.

---

* This was coined by Paul Ehrenfest in 1911, in reference to the Rayleigh-Jeans model from 1905, and would be a great name for a band.

At low frequency, then, there are relatively few oscillators, because there are few possible standing waves at relatively long wavelengths, but each emits many "quanta" worth of light. At high frequency, there are many oscillators (because there are many allowed modes at shorter wavelengths), but each emits little or no light. The competition between the increasing number and decreasing emission gives exactly the kind of peaked spectrum observed in black-body radiation: starting at long wavelengths and moving down, the increase in the number of oscillators is initially faster than the decrease in light emitted per oscillator, so the total amount of light increases to a peak, then decreases as the emission cuts off completely. And it explains the shifting peak of the spectrum, as well: as the temperature increases, the amount of heat energy increases, increasing the share allotted to each mode, and pushing up the frequency where the quantum hypothesis cuts off the light emission.

Planck initially introduced the quantum hypothesis thinking it was a "desperate mathematical trick." And in fact, it *was* a bookkeeping trick of a type often employed in calculus. Mathematical physicists regularly describe smooth, continuous phenomena in terms of discrete steps when setting up a problem, then use well-honed mathematical techniques to make the "steps" infinitesimally small and restore the original smoothness. Planck knew that giving each oscillator a characteristic energy that increased with frequency would give the resulting spectrum the cutoff he needed, but he also thought he would be able to use calculus to reduce the constant multiplying the frequency to zero, restoring the smoothness and doing away with the steplike quanta of energy. Instead, he found that the constant needed to take a very small but stubbornly nonzero value: these days, it's called "Planck's constant" in his honor, and goes by the symbol h, with a value of 0.0000000000000000000000000000000006626 joule-seconds—a very small number indeed.* With the quantum hypothesis in place—namely that energy comes in discrete, irreducible "packets"—and h taking that tiny but nonzero value, the process of dividing the available energy

---

*   It is more commonly written as $h = 6.626070040 \times 10^{-34}$ kg–m²/s.

among all the possible frequencies leads to exactly the formula that Planck had found to describe the black-body spectrum.

Planck's formula is a spectacular success, and has become an invaluable tool for many areas of physics. Astronomers use it to determine the temperature of distant stars and gas clouds by measuring the spectrum of the light they emit. The spectrum of light from a typical star—our sun included—closely resembles a black-body spectrum, and by comparing the light we see to the prediction from Planck's formula, we can deduce the temperature on the surface of stars many light-years away.

Probably the most perfect black-body spectrum ever measured is the "cosmic microwave background" mentioned earlier, a weak radiation field in the radio-frequency part of the spectrum that permeates the entire universe. This background radiation is one of the best pieces of evidence for Big Bang cosmology: the microwaves we see today were created about 300,000 years after the Big Bang, when the universe was still extremely hot and dense, but had cooled just enough to allow photons to escape. Over the intervening billions of years, the universe has expanded and cooled, so what were once high-energy, visible-light photons reflecting a temperature of thousands of kelvin have been stretched out to microwave wavelengths. The spectrum has been measured many times, and matches a black body at about 2.7K to phenomenal accuracy. In fact, tiny variations in the temperature of that background radiation from different points of the sky—shifts of millionths of a kelvin—provide the best information we have about the conditions of the very early universe, and the origins of galaxies, stars, and planets.

On a more down-to-earth level, the Planck formula informs the way we talk about light and heat every day. Photographers and designers talk about the "color temperature" of various kinds of light, which is a number in kelvin that corresponds to the temperature of the black body whose visible spectrum most closely matches the light in question.* The different styles of lightbulbs available at your favorite

---

*   Human perception makes the language around color and temperature confusing: reddish light is traditionally called "warm," even though it corresponds to a lower temperature source, while bluish light is called "cool."

home-improvement store—"soft white," "natural light," and so on—use a variety of techniques to produce light with a spectrum that resembles black-body radiation from objects of different temperatures.

In the context of breakfast, black-body radiation can be used to determine the temperature of hot objects—if your kitchen contains one of those infrared thermometers that you point at a pan to see whether it's hot enough, you're making use of Planck's formula. A sensor in the thermometer detects the total amount of invisible infrared radiation coming from whatever it's pointed at, and uses that to deduce the temperature of a black body that would emit that much infrared light.

Despite the many successes of his formula and the personal fame it brought him, Max Planck himself was never particularly satisfied with his quantum theory. He regarded the quantum hypothesis as an ugly ad hoc trick, and he hoped that someone would find a way to get from basic physical principles to his formula for the spectrum without resorting to that quantum business. Once the idea was out there, though, other physicists picked it up and ran with it—most notably a certain patent clerk in Switzerland—leading to a complete and radical transformation of all of physics.

CHAPTER 3

DIGITAL PHOTOS: THE PATENT CLERK'S HEURISTIC

My social media feeds are full of the usual overnight fare—morning news from Europe and Africa, evening stories from Asia and Australia, **digital photos** *of the kids and cats of friends around the world . . .*

As someone who writes regularly about historical discoveries in science, I'm often struck by how few photos exist of major scientists of the past. These image collections also tend to be skewed toward later life, *after* the subject became famous, which distorts our perception of scientists somewhat. Photos of Einstein taken around the time he was revolutionizing physics show a well-groomed young man—a far cry from the iconic images of him taken later on, with rumpled clothes and wild white hair. The scarcity of images is complicated by copyright issues, of course, but even professional archives tend to have only a few dozen photos of great physicists of the twentieth century.

This paltry number is especially shocking from a modern perspective; over the last few decades digital photography has become ubiquitous, leading to an explosion in the number of images created. I've long been interested in photography, but the expense of purchasing film and having it developed presented enough of an obstacle that I have only a few hundred pictures from before 2004, when I first got a digital camera. Since then, I've taken tens of *thousands* of digital photos, nearly all of which I have stored on the hard drive of my computer. I probably have more photos of my children (who'll be ages ten and seven when this book comes out) than have been taken of my parents in their entire lives. And that only counts those captured with my dedicated camera, not quick snapshots grabbed with my phone.

The incredible ease of digital photography, particularly thanks to the spread of cameras in phones, has had a revolutionary impact on everyday life. Today there are billion-dollar companies that do nothing but process, store, and share photos taken by users, and whole new cultural phenomena like "selfies" that have grown up around the technology. And the ready availability of cameras has transformed all manner of interactions between the general public and various authority figures. Incidents that would have been "he-said, she-said" disputes in the days of film seem invariably to be caught on cell phone video these days, with consequences for society that are still working themselves out.

Digital cameras made the transition from rare and expensive gadgets to integral parts of everyday life impressively quickly, but the science underlying these devices remains underappreciated. The sensor your phone uses to take pictures of your kids, cats, or breakfast to post on Twitter is, at a fundamental level, quantum mechanical, relying on the particle nature of light. There's no small irony, then, in the fact that the discovery of the physics essential to this technology was merely a byproduct of an experiment proving light's *wave* nature.

HERTZ'S EXPERIMENTS

As mentioned in the last chapter, experiments by Thomas Young and François Arago in the early 1800s—demonstrating that light waves show interference effects when passing around obstacles—conclusively showed that light behaved like a wave. And in the middle of that century, Maxwell's equations answered the question "What is waving?" by predicting the existence of electromagnetic waves moving at the speed of light.

One of the implications of a theory of light as an electromagnetic wave is that it ought to be possible to create such waves using electric currents. In the late 1880s, the young German physicist Heinrich Hertz decided to do just that, and put Maxwell's equations to a direct experimental test. Hertz devised an ingenious apparatus involving "spark gaps," pairs of metal knobs separated by a few millimeters of air. One spark gap was attached to an antenna connected to a battery system, which applied an oscillating high voltage between the knobs. This produced a bright spark in the gap as the electric field broke down the air between the knobs, allowing current to flow at a frequency determined by the oscillating voltage (which Hertz could set to a value of his choosing). As electrons rushed back and forth across the gap, according to Maxwell's equations, their motion should generate electromagnetic waves traveling outward from the gap, oscillating at the same frequency.

The other spark gap—the knobs on either end of a wire ring placed some distance away—served as the detector. The arriving electromagnetic wave from the transmitting spark gap induced a smaller voltage in the detector and produced a much smaller spark. The distance between the knobs on the detector was adjustable, and would be tuned until the arriving waves just barely created a spark across the gap. More intense arriving waves induced a higher voltage in the detector, increasing the distance the spark could jump. Using this detector, Hertz was able to map out the intensity of the waves produced and show that the results exactly matched Maxwell's predictions—both for traveling waves leaving the detector and standing waves formed by reflecting the initial waves off a metal sheet on the far side of a lecture hall. Hertz's apparatus generated waves at extremely low frequencies compared to visible light, but he showed that they traveled at the same speed, confirming that light is an electromagnetic phenomenon.

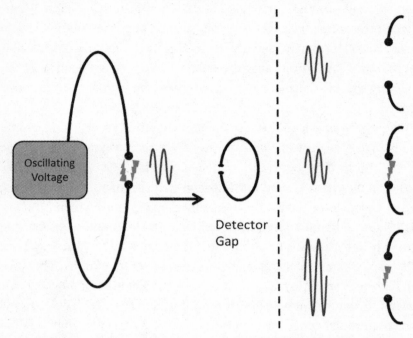

Principle of the spark gap apparatus used by Hertz. A large oscillating voltage creates sparks across a gap in a loop of wire, generating electromagnetic waves at that frequency. At the detector gap, the wave induces a voltage that can make a spark if the gap is small or the wave is large enough. The size of the largest gap the spark can jump is a measure of the size of the wave.

Asked about the significance of his experiments, Hertz demonstrated the business acumen of a great physicist by cheerfully responding, "It's of no use whatsoever. This is just an experiment that proves Maestro Maxwell was right—we just have these mysterious electromagnetic waves that we cannot see with the naked eye. But they are there." Only a few years later, however, the same principles used in Hertz's spark-gap experiment were used to generate radio waves for "wireless telegraphy," eventually leading to broadcast radio, television, and cellular phones.

Hertz's experiment demanded enormous care and precision, plus the investigation of many possible confounding factors. In the course of this investigation, Hertz noticed that the size of the detector gap possible for a given configuration was slightly larger when there was a direct line of sight from the source to the detector. When he blocked the light from the initial spark from falling on the detector, it reduced the size of the gap across which a spark could jump. He thus discovered what's now called the "photoelectric effect": that ultraviolet light falling on a metal surface produces a charge on the metal. This charge makes it easier for weak incoming waves to induce a spark between the knobs of his detector.

For Hertz, the discovery of the photoelectric effect was of little consequence, merely a systematic quirk to be explained along the way to demonstrating the wave nature of light. Unbeknownst to him, though,* this minor digression would prove to be an essential bit of evidence for light's *particle* nature just a few decades later.

THE PATENT CLERK'S HEURISTIC

Hertz's accidental discovery of the photoelectric effect drew the attention of a number of prominent physicists of the day, who began shining ultraviolet light on a variety of materials and investigating the results. From the way the ejected particles responded to electric and magnetic fields, they determined that the charges ejected by the light were electrons, which had recently been identified as negatively charged subatomic particles by the British physicist J. J. Thomson (who would eventually win the 1906 Nobel Prize for discovering the electron).

When combined with the wave model of light, the knowledge that the photoelectric effect involves the ejection of electrons, which are components of atoms, allowed physicists to construct an appealingly simple model of the process. Electrons are bound into atoms, and an

---

* Around five years after his pioneering experiments on electromagnetic radiation, Hertz died of a vascular disease at only thirty-six, a tragic loss for physics.

incoming electromagnetic wave shakes those electrons back and forth. This shaking transfers energy to the electrons in a way that physicists expected would depend on the intensity of the light. The higher the intensity, the greater the displacement of electrons, so high-intensity light should deposit enough energy to knock electrons loose quickly, but as electrons will continue to absorb energy as long as the shaking continues, even low-intensity light should eventually shake a few electrons free.

The light's frequency was another experimental factor that could affect the ejected electrons, though how the electron properties should depend on frequency was less obvious. In the classical wave picture of light, the amount of energy carried by the wave depends on the *size* of the wave, not its frequency, so any dependence on frequency would be more complex than the intensity dependence. There might be some resonance effects—shaking at some characteristic frequency associated with a particular atom might deposit energy more efficiently, in the same way that gentle shaking of a pendulum at just the right rate can produce dramatic oscillations. Lower frequency might also lead to a delay in emission, as electrons shouldn't be ejected until they've had time to be shaken back and forth a few times, but the frequency of visible light is so high there's no real hope of measuring this.

The simple model favored by physicists made four basic predictions about the behavior of the ejected electrons that can be tested in experiments:

- First, the number of electrons emitted should increase as the intensity increases—the harder you shake the electrons inside individual atoms, the more electrons should come out.
- Second, the energy of the electrons coming out of the material should increase with the intensity—when you shake them harder, electrons should come out moving faster.
- Third, there should be some delay in the emission of electrons, especially at lower intensity and lower frequency—dim light and slow shaking should take some time to build up enough energy to pop an electron loose.

- And finally, the number and energy of the electrons emitted should, if they depend on the frequency of the light at all, show some resonant behavior.

This simple model ties together the best knowledge of the day concerning the physics of light and electrons, and thus was very appealing to physicists. Unfortunately, it was also a miserable failure.

In particular, careful experiments by the German physicist Philipp Lenard (who had worked with Hertz for a time) failed to show the expected relationship between the intensity of the light and the energy of the electrons. Brighter light did, as expected, increase the *number* of electrons emitted (measured by the current flowing between two metal plates inside a vacuum tube when one of the plates is illuminated), but the energy of those electrons (measured by the voltage associated with the current in the vacuum tube) was the same regardless of the intensity of the light used.

An even more puzzling outcome of Lenard's experiment was the discovery of a surprisingly simple relationship between the energy of the ejected electrons and the *frequency* of the light. Across all the materials Lenard tested, the energy of the electrons increased as the frequency increased, in an apparently linear manner. This was a completely unexpected and deeply mysterious result.

As in the case of thermal radiation, the simple and universal behavior discovered by Lenard seemed to point to simple underlying physics, but nobody could construct a convincing model. Lenard himself spent many years working on the theory that the electron energy was determined by the motion of electrons within the atoms, with the light serving only as a trigger for the electron ejection, but this proved untenable and he eventually had to abandon it.

The explanation that would eventually become the accepted model for the photoelectric effect was first proposed in 1905 by an obscure patent clerk in Switzerland by the name of Albert Einstein. In a paper with the rather cautious title "On a Heuristic Viewpoint Concerning the Production and Transformation of Light," Einstein suggested taking Max Planck's quantum hypothesis, which associated each light-emitting "oscillator" in a material with a characteristic energy that depended on the frequency of its emitted light, and applying it to the light itself. In this "heuristic viewpoint," a beam of light is not a wave, but a stream of particles (now called "photons," though that term wasn't coined until years later; Einstein preferred "light quanta"), each carrying a single quantum of energy: Planck's constant multiplied by the frequency of the light. If the energy of a single photon exceeds a characteristic energy for the material being illuminated, called the "work function," each photon can knock loose a single electron, which carries off the rest of the photon's energy.

This particle model of light was a radical departure from well-known physics, but it worked brilliantly to explain the observed features of the photoelectric effect. A more intense beam of light contains more photons, thus enabling the increase in the number of emitted electrons. The energy of the electrons, though, does not depend on the intensity, because only a single photon is needed to knock loose an electron. And the increase in energy with increasing frequency simply reflects the increasing energy of a single photon following Planck's rule relating energy to frequency; if the photon energy is greater than the work function, the electron carries off the excess, which increases as the frequency increases.

Einstein's photon model is simple and elegant, but also completely incompatible with Maxwell's equations—which only work for waves, not particles—and was thus wildly unpopular when first introduced. Planck himself, in nominating Einstein to the Prussian Academy of Sciences, wrote: "That he may sometimes have missed the target in his speculations, as for example, in his hypothesis of light quanta, cannot really be held too much against him, for it is not possible to introduce fundamentally new ideas, even in the most exact sciences, without occasionally taking a risk."

However unpopular it was, Einstein's heuristic model made very clear and unambiguous predictions about what one should expect to see in photoelectric-effect experiments, and as a result attracted a considerable amount of attention. The situation remained a little murky, though, until Robert Millikan, one of the finest experimental physicists of the day, took up the question. The experiments are very sensitive to contamination of the metal surfaces and small voltage shifts arising from contacts between different metals, but Millikan and his team[*] tracked down and resolved all these issues, and provided an extremely convincing experimental confirmation of Einstein's model in 1916, producing a measurement of Planck's constant that was consistent with previous values but with much improved precision.

This does not mean, however, that Millikan was a fan of the photon model. In fact, the introduction of his first paper on the subject is a masterpiece of the passive-aggressive style in scientific writing:

> *Einstein's photoelectric equation for the maximum energy of emission of a negative electron under the influence of ultra-violet light . . . cannot in my judgment be looked upon at present as resting upon any sort of a satisfactory theoretical foundation. Its credentials are thus far purely empirical . . .*
>
> *I have in recent years been subjecting this equation to some searching experimental tests from a variety of viewpoints, and have been led to the conclusion that, whatever its origin, it actually represents very accurately the behavior . . . of all the substances with which I have worked.*

Millikan's grudging acceptance of the accuracy of Einstein's model in spite of his personal reservations is fairly representative of opinion at the time. The photon model was too radical a departure to be easily accepted, but it worked too well to be easily cast aside. Over time, the particle view of light became more accepted, though concerted efforts to find an alternative explanation continued until the mid-1920s. In a

---

[*] As was common in that era, Millikan is listed as the sole author on the resulting papers. His acknowledgments, however, make clear that other people (he credits A. E. Hennings and W. H. Kadisch for assistance with the experiment, and he thanks Walter Whitney for spectroscopic measurements to determine light wavelengths) contributed at a level that by modern standards would rate an author credit. He also gives a rather generous acknowledgment to "the mechanician, Mr. Julius Pearson," for helping design and make the evacuated glass tubes used for the experiment.

strict technical sense, incontrovertible experimental proof of the existence of photons was only achieved in 1977,* but as a practical matter,
light as a particle was an accepted part of quantum physics by 1930 or
so.

Both Einstein and Millikan made out well as a result of the photoelectric effect. While he's best known for relativity, the photoelectric
effect is the only specific result mentioned in Einstein's citation for
the 1921 Nobel Prize in Physics.† And Millikan's own Nobel Prize, in
1923, mentions both the photoelectric effect and an earlier experiment
to measure the charge on an electron. And as we shall see, this new
understanding of light paved the way for many technologies that have
become central to modern life.

PHOTOELECTRIC TECHNOLOGIES

The dual particle and wave nature of light is one of the classic examples
of the weirdness of quantum physics, a phenomenon with seemingly
contradictory properties. This is evident in the photoelectric effect
itself, which relates a particle property (the energy content of a single
photon) to a wave property (the frequency of the light), leading to
some potential confusion as to what, exactly, it means for a particle to
have a frequency. Even today, physicists continue to argue about the
best language to describe the nature of light, and how best to teach the
core concepts.

As such, the idea of photons may seem too bizarre to make everyday use of. In fact, however, it is central to essentially any technology
used to convert light into an electronic signal.

---

*   In the 1960s, Leonard Mandel and colleagues developed a "semi-classical"
     model for the photoelectric effect, where the metal surface is treated quantum-
     mechanically but the light is considered as a classical wave. In 1977, an
     experiment by Jeff Kimble, Mario Dagenais, and Mandel demonstrated a clear
     delay between the emission of consecutive photons by single atoms, an effect
     that can only be explained with a particle model.
†   This came about through tedious and petty academic politics.

Admittedly, the device that shows the clearest connection to photo-electric physics is a bit arcane: it's what's known as a "photomultiplier tube," consisting of a series of metal plates with a high voltage (generally a few hundred to a thousand volts) applied between them. A photon of light falling on the first of these will eject a single electron through the photoelectric effect. The high voltage then accelerates this electron toward the next plate in the series, where it collides and knocks loose several more (ten to twenty) electrons.* Each of these is then accelerated toward the next plate, and the next, and so on. By the end of the tube, a single photon has triggered a cascade of millions of electrons, producing a tiny pulse of current that can readily be detected. Photomultiplier tubes can be extremely sensitive, able to detect even a single photon, and they are at the heart of many experiments investigating the quantum nature of light. While used in some older "electric eye" systems, these days, photomultiplier tubes are generally found only in physics labs.

The same essential physics, however, is at the heart of a digital camera. Each pixel in a digital camera's sensor consists of a tiny chunk of semiconductor material that is exposed to light for some period of time. In this case, incoming photons do not completely eject electrons from the material, but they do promote an electron from a state in which it is unable to move to one where it can flow freely (more about this in Chapter 8). When the camera shutter is open to take a photo, all the electrons within a given pixel that are promoted to a freely flowing state are collected,† building up a voltage that gives a measure of the brightness of the light hitting that pixel. At the end of the exposure time, these pixel voltages are read out to produce an image.

---

*   As a material particle with mass and charge, an electron colliding with a surface delivers energy to the material more effectively than a massless photon does.

†   In an older "CCD"-type camera, the electrons build up in each pixel, and after the exposure is finished, they are shifted along the rows of pixels to a sensor at the edge of the chip. The "CMOS" sensors on most newer cameras include a small amplifier associated with each pixel, and directly produce a voltage signal that's read out to make the image.

Silicon-based photosensors offer the great advantage of small size and ready integration with the digital information processors they work alongside. Today, a camera chip small enough for use in a cell phone will contain a number of pixels that rivals the resolution of a professional-quality digital camera. The camera of my current smartphone has 16.1 million pixels (the default image is 5344 × 3006 pixels), while my good DSLR camera has 24 million (6000 × 4000). The primary limitation on the quality of cell phone photography these days is optical, not electronic: a lens package small enough to incorporate into a phone has more limited capabilities than the larger lenses of a standalone camera. For most people who are not serious photography buffs, though, these limits are not particularly noticeable.

To make color sensors, a grid of red, green, and blue filters is placed over the top of the pixel array, so that each pixel is detecting light of a single color. To make the final image, the voltages from nearby pixels of different colors are combined to determine the mix of red, green, and blue colors that best approximates the light at that point in the image.

Digital cameras can get away with measuring only three colors because this closely matches the way the human eye processes light to determine color. When a photon strikes a light-sensitive cell in the retina, the energy from the photon triggers a change in the configuration of a protein molecule, which sets off a chain of chemical reactions that eventually sends a signal to the brain to inform it that this particular cell detected some light. There are three varieties of these cells, each sensitive to a different range of photon wavelengths, and the brain uses the different responses from each type to produce the color that we see. The peak sensitivities are at wavelengths corresponding to blue, green, and yellow-green light, though all three are sensitive to a broad range. Our brains infer color from the mix of activity levels of these cells: red light triggers only the longest-wavelength receptors and blue light only the shortest, while green light triggers all three. Televisions and computer monitors use a mix of the three colors to trigger these receptors in the right proportions to duplicate our response to the spectrum of light from some real-world object and trick the brain into thinking it sees a rich variety of colors.

While it takes only a single photon to trigger the light-detecting process, a typical digital camera sensor does not offer single-photon sensitivity because the random thermal motion present in any material at temperatures above absolute zero can spontaneously generate free electrons inside the sensor. To have confidence that the signal recorded by a particular pixel indicates actual light, the number of photoelectrons must exceed this "dark current" to register a response within the sensor, which limits the sensitivity at low light. This effect depends strongly on temperature, so professional scientific cameras used by astronomers and in quantum-optics experiments generally have their sensors cooled to reduce the dark current to a level that allows reliable detection of single photons.

The same issue of dark current affects your eyes—in principle, the photosensitive chemicals in your retina can detect a single photon, and in carefully controlled laboratory experiments, human volunteers can sometimes detect pulses of light containing only a handful of photons. In more typical situations, though, it takes something like one hundred photons entering the eye within a few milliseconds for a human to reliably detect a faint flash of light—and, of course, cooling the retina of the human eye to reduce dark current and improve sensitivity isn't advisable.

However, the limits of dark current are a practical issue, not a fundamental one. The process that enables commercial digital cameras is fundamentally quantum: a single photon enters the sensor and knocks loose a single electron. Our ability to understand this process—and to build these devices that take advantage of it—traces directly back to Heinrich Hertz's chance discovery of the photoelectric effect, and Albert Einstein's radical suggestion in 1905 that light might be a particle after all.

CHAPTER 4

THE ALARM CLOCK: THE FOOTBALL PLAYER'S ATOM

*The sun comes up not long before **my alarm clock** starts beeping, and I get out of bed to start the day . . .*

In a strict technical sense, a new day begins when the sun rises, but as a practical matter, *my* day begins when my alarm clock beeps. These two events are generally closer together than I would prefer, and for much of the winter they're not in the right order, but while the sun may start the astronomical day, it's the clock that marks the start of the working one.

The specific timepiece on my nightstand is nothing all that special—a cheap plug-in digital clock with few features other than a shrill beeping alarm sufficiently irritating to wake me from a sound sleep. The modern accounting of time that it embodies, however, is deeply rooted in the quantum physics of atoms and the wave nature of material objects.

This is just the latest step in a long chain of time-measuring technologies stretching back to prehistory.

A BRIEF HISTORY OF TIMEKEEPING

The measurement of time is arguably the field of technology with the longest documented history, stretching back to the days before written language. The passage tomb at Newgrange in Ireland, an artificial hill constructed around 3000 BCE from 100,000 tons of earth and rock, is in fact a sophisticated timekeeping device. Inside the hill, a twenty-meter passage leads to a vaulted chamber in the center. This central chamber remains dark all year, save for a few days around the winter solstice, when the rising sun casts a ray of light through a small opening above the door all the way down the passage. This provides an unambiguous way of marking the turning of the year, and it still functions perfectly some five thousand years after the structure was built.

The science and technology of time have come a long way since Newgrange's era, but the fundamental principle remains the same: we mark the passage of time by counting occurrences of some regular, repeated event. For a calendar marker like Newgrange, the regular, repeated motion is the shifting position of the rising sun over the course of the year, which (in the northern hemisphere) rises north of due east during the summer months, and south of due east in the winter. The winter solstice is the shortest day of the year and the southernmost position of the rising sun, an extremely reliable pattern that Newgrange's builders must have observed over many years before building their giant monument.

Astronomical motions can be used to track time in shorter intervals as well, for instance by using a sundial: the direction of the shadow cast by a vertical object indicates the time of day. At night, the apparent motion of the stars across the sky works in much the same way. Both of these timekeeping methods are complicated somewhat by the orbital motion of the earth, but because these patterns have been closely tracked for thousands of years, it's possible to keep fairly accurate time watching only the sun and stars.

Of course, using astronomical observations for timekeeping has its limits: it requires clear skies, which can't always be relied upon, and it's difficult to use sundials or star positions to time anything taking less than several minutes. For shorter time scales and bad-weather operation, timekeepers turned to objects relying on the regular motion of some substance. Water clocks (where a time interval is defined by the emptying of a vessel) were used in ancient Egypt and China, and sand timers were invented in medieval Europe, where water clocks were prone to freezing in the winter.

For settled agrarian societies, these methods may be sufficient, but with the rise of globe-spanning empires in the 1500s and 1600s, a need for ever more accurate timekeeping arose. Navigators crossing oceans, out of sight of land for weeks at a time, need to know both latitude and longitude to track their position on a map. Latitude can readily be determined by the height of the sun at noon, but accurately measuring a change in longitude requires knowing the time not just where you have ended up, but back at your starting point as well. Improved astronomical tables provided one method of tracking the passage of time and thus longitude, but portable mechanical clocks that marked time by the motion of a swinging pendulum or oscillating spring made this easier still. Making a mechanical clock that can keep time through an ocean voyage was a formidable technical challenge,* but by the mid-1800s, such clocks were in regular use. These too, though, were accurate only to a point, and the rise of continent-spanning networks of railroads and telegraphs only accelerated the drive for timekeeping precision.

The problem faced by scientists studying time is that any clock based on the motion of physical objects is inherently unreliable. Mechanical clocks are sensitive to small differences in their manufacture: tiny variations in the shape of two pendulums will cause their respective clocks to tick at slightly different rates. Even astronomical clocks are prone to drift: the rotation of the earth is slowing down over time due to the gravitational influence of the moon, which is why every

---

* Part of this story is told in Dava Sobel's award-winning book *Longitude*.

few years you'll hear news stories about a "leap second" being added at midnight on December 31.

The ideal clock would be one with no physical moving parts—and with the realization that light is an electromagnetic wave, such a clock became possible. A light wave is an electric field that oscillates back and forth at some frequency, and once that field is set in motion, it is extremely difficult to change the frequency of the oscillation.* If we can count these oscillations, then, we can use the light as a clock.

The chief obstacle to the use of light to measure time is finding a way to generate light whose frequency is known absolutely. It's not difficult to generate waves at a single frequency (that is, not a broad spectrum like the black-body radiation from a hot object) by driving electric currents, as demonstrated by Hertz's experiments discussed in the previous chapter. However, the exact frequency of those oscillating currents depends strongly on the physical circuit used to make them, which leaves us with the same problem presented by mechanical clocks using springs and pendulums, namely the difficulty of making two truly identical objects. What's more, to build a high-precision clock based on light, we need a way to make light with a frequency that is not only known, but will be exactly the same no matter where and when the clock is operated.

The solution to this problem emerged from a seemingly unrelated problem, a mystery involving the way that light interacts with individual atoms.

THE MYSTERY OF SPECTRAL LINES

For many years, the study of atoms developed more or less independently from the study of the nature of light. The two are powerfully

---

* Because the speed of light is different in different media, the wavelength of light will change as it moves from one medium to another—from air into glass, for example—while the frequency of oscillation remains the same.

linked, however, because light is the principal tool used to discover the internal structure of atoms.

In the early 1800s, around the time Arago was conclusively proving the wave nature of light, other physicists were making discoveries about the light emitted by different substances. William Hyde Wollaston noticed some dark "lines" in the spectrum of the sun. Sunlight that is passed through a vertical slit and then dispersed with a prism produces a broad band of colors, but in certain narrow ranges, there's dramatically less light than at frequencies just a bit higher or lower. Wollaston initially tried to interpret these as boundaries between the discrete colors of the spectrum—the "ROY G BIV" sequence you learn in grade school—but there were too many lines, and in the wrong places. The "boundary" model was completely destroyed in 1814 when Joseph von Fraunhofer obtained more accurate spectra using a diffraction grating—which relies on the wave interference of light to disperse the different wavelengths—and identified several *hundred* dark lines in the solar spectrum. Fraunhofer launched the systematic study of these lines, determining their wavelengths and classifying them based on intensity. These dark lines in the solar spectrum are now referred to as "Fraunhofer lines" in his honor, recognizing his contributions to launching the field of spectroscopy.

Around the same time as Fraunhofer's observation of dark lines in the spectrum of the sun, other scientists, notably William Henry Fox Talbot and John Herschel, noticed the presence of *bright* lines in the spectrum of light emitted by various chemical compounds when heated in a flame. These flame spectra come from minute quantities of material vaporized in the heating process, and such diffuse vapors produce a very different spectrum than the thermal radiation from a large hot object. Where the spectrum of the black-body radiation that Planck would explain at the end of the century depends only on the temperature, the flame spectra depend very sensitively on exactly what element is being heated: each element emits light only in very narrow lines at particular wavelengths. In fact, Talbot and Herschel showed that these bright lines could be a useful tool for identifying minute quantities of particular elements. The French physicist Jean Bernard Léon Foucault demonstrated that a relatively cool vapor of a given element would absorb light at the same wavelengths emitted by that element when heated in a flame. This provides a conceptual explanation of Fraunhofer's dark lines: the "missing" light from the sun's spectrum is that emitted by the hot center of the sun and then absorbed by elements in the cooler outer layers of the solar atmosphere.

The disparate spectroscopic investigations of the early 1800s were brought together in a systematic way in the 1850s through the work of Gustav Kirchhoff and Robert Bunsen, who established spectroscopy as a subdiscipline of physics with formal rules and procedures. Kirchhoff and Bunsen showed that every known chemical element produced a *unique* pattern of spectral lines, in both emission and absorption. Within only a few years, spectral lines were being used to identify new elements. The most spectacular example of spectroscopic discovery is that of helium, which was identified in 1870 based on a new spectral line found in light from the sun—a narrow region at a wavelength of 587.49 nm (in the yellow part of the spectrum) with much more light than the black-body–like spectrum to either side of it—but was not isolated on Earth until the 1890s.

These spectral lines provide the conceptual foundation for a clock based on light: if each element emits and absorbs only very specific frequencies of light, we can obtain a known frequency of light for use in a clock by selecting a particular spectral line of a particular element. For this to have any real appeal, though, would require that physicists understand how atoms produce those spectral lines and how their frequencies are determined from the laws of physics, in order to be absolutely confident that the frequency is reliable. While Kirchhoff and Bunsen established the existence of spectral lines as an empirical fact and a useful tool for physics and chemistry, the origin of these lines remained a mystery.

This proved a difficult problem to crack, as the spectra of many elements are very complex, with large numbers of lines through the visible spectrum, and identifying useful patterns in these forests of lines was tricky. The spectrum from the lightest element, hydrogen, finally provided the clue that cracked the case. Hydrogen's visible spectrum consists of only four lines, at wavelengths of 656, 486, 434, and 410 nanometers. The simplicity of this spectrum seemed to hint at a simple underlying principle, and in 1885 a Swiss mathematician and school-teacher, Johann Balmer, found that if he assigned integer numbers to the visible lines of hydrogen (3, 4, 5, and 6 respectively), he could accurately predict their wavelengths using a simple mathematical formula. A few years later, the Swedish physicist Johannes Rydberg extended Balmer's work, associating all of the spectral lines in hydrogen (the visible lines used by Balmer, and similar series of lines in the ultraviolet and infrared regions) with *pairs* of integers: one, called *m*, to identify the particular region of the spectrum (1 for the "Lyman series" in the ultraviolet, 2 for the visible Balmer lines, and 3 for the "Paschen series" in the infrared), and the other, *n*, a line within that series. In modern notation, Rydberg's formula for identifying the wavelengths of these lines—wavelength being traditionally represented by the Greek letter lambda (λ)—looks like this:

$$1/\lambda = R\left(\frac{1}{m^2} - \frac{1}{n^2}\right)$$

The symbol R stands for a constant, now known as the "Rydberg constant" with a modern value of 10,973,731.6 "inverse meters," or 1/m (to match the "$1/\lambda$" on the other side), whose value determines all the wavelengths emitted by hydrogen.

Rydberg's formula worked very nicely to explain the wavelengths of all known spectral lines in hydrogen, and with some small tweaks can explain some series of lines seen in other elements. Rydberg's formula may not have worked for all elements, but it was the only successful system anybody had managed to come up with, and its mathematical simplicity seemed to hint at some similarly elegant underlying structure. Unfortunately, for the next twenty-five years, nobody had any idea what that structure could be.

THE MOST INCREDIBLE THING: INSIDE THE ATOM

The real breakthrough in explaining the light emitted and absorbed by hydrogen, and eventually all the other elements, came in 1913, the work of the Danish theoretical physicist Niels Bohr. It was preceded by another shocking discovery, though, made in the lab of Ernest Rutherford in Manchester, England.

In 1909, Rutherford was already established as a major force in physics, having just been awarded the 1908 Nobel Prize in Chemistry, for research carried out at McGill University in Montreal between 1898 and 1907. This work gave us the classification of radioactivity in terms of "alpha," "beta," and "gamma" emission, still in use today; showed that alpha particles were helium nuclei (we'll talk more about alpha decay in Chapter 10); and demonstrated that it was their emission that changed one chemical element into another. This discovery regarding change of chemical identity is the reason Rutherford's Nobel is in chemistry, a fact that is not without irony. Rutherford was famously disdainful of sciences other than physics, reportedly declaring that physics was the only real science, and "all the rest is stamp collecting." He made light of this in his remarks at the Nobel banquet, joking that of all the transformations he had studied, none was more rapid or surprising than his own change from physicist to chemist at the instant of winning the prize.

Never one to rest on his laurels, Rutherford launched a new program of research on moving to Manchester in 1907. The idea was to direct alpha particles produced by the radioactive decay of radium toward a piece of gold foil, and use the way the paths of these particles were deflected as they passed through the foil to infer some details about the structure of matter. The best atomic model at that time was J. J. Thompson's "plum pudding" model, which pictured the atom as a blob of positive charge filling the whole volume, with negatively charged electrons embedded in it. Such an atom would only weakly resist the passage of the high-energy alpha particles from Rutherford's source, deflecting them from their course by a tiny amount, a few degrees at most. Early experiments looking for alpha particles at these small deflections showed mostly what scientists expected. As a sanity check on these results, though, Rutherford set his research assistant Hans Geiger and an undergraduate student named Ernest Marsden to the task of checking for alpha particles deflected by more than 90 degrees, leaving them on the same side of the foil as the radioactive source.

While the prevailing theory said they should find none, Marsden and Geiger in fact found substantial numbers of alpha particles deflected by large angles—up to 150 degrees, nearly straight back at the source. To call this unexpected is an understatement; Rutherford himself, some years later, said:

> It was quite the most incredible event that has ever happened to me in my life. It was almost as incredible as if you fired a 15-inch shell at a piece of tissue paper and it came back and hit you."

According to the "plum pudding" model of the atom, the large deflection angles measured by Marsden and Geiger were simply impossible. The electrostatic repulsion between a high-energy alpha particle and a diffuse ball of positive charge like that in the "plum pudding" gold atoms making up the foil could simply never be strong enough to make an alpha particle reverse course

Rutherford recognized this almost immediately, and he realized that Marsden and Geiger's shocking result could be explained only if the atom's positive charge was not diffuse but concentrated—that is, if a positively charged core contained the vast majority of the atom's mass. Rutherford's proposal was the birth of the modern cartoon version of an atom, featuring a tiny, positive nucleus orbited by negatively charged electrons.

Based on this assumption that most of the mass of the atom lay in the tiny nucleus, Rutherford worked out an equation to predict how the number of alpha particles that were deflected to a particular angle should depend on the energy of the alpha particles and the composition of the target. Marsden and Geiger carried out a new series of experiments, which confirmed all the predictions of Rutherford's scattering formula.

As with Einstein's photoelectric model in the previous chapter, though, the manifest empirical success of Rutherford's model did not immediately lead to its widespread adoption. The reason for this is simple: according to well-understood classical physics, Rutherford's atomic model is impossible. An electron in orbit around the nucleus will be constantly changing its direction of motion, which means it's accelerating—and that acceleration should lead to the rapid death of a Rutherford atom. Accelerating charges *radiate*: this is the principle used by Hertz to generate electromagnetic waves for his experiments, and for every radio transmitter built in the last century and a half. An orbiting electron should spray out high-frequency light waves—x-rays and gamma rays—in all directions, and those waves should carry away energy, causing the electron to slow down and spiral inward until it crashes into the nucleus. Rutherford's solar-system atom is simply absurd, from the standpoint of classical physics.

ENTER THE QUANTUM

So, Rutherford's model of an atom with most of its mass in the nucleus worked very well to explain the scattering experiments done by Marsden and Geiger, but thanks to the fundamental conflict between the notion of orbiting electrons and classical physics, it wasn't taken all that seriously outside of Manchester. Happily, Niels Bohr was about to arrive to spend a few months working with Rutherford, and he would end up cracking the problem and transforming our understanding of the atom.

Bohr and Rutherford made an odd pairing; Bohr was notoriously equivocal and soft-spoken, while Rutherford was a forceful presence with a booming voice. (Once, when Rutherford was giving an interview to a US radio program, a colleague came looking for him. Informed that Professor Rutherford was speaking to America via the radio, the visitor responded, "Why does he need the radio?") The contrast between Bohr and Rutherford carried over to their work: while Rutherford was quite capable mathematically, he often disparaged pure theory, and Bohr was very much a theorist. Teased about his decision to bring Bohr in, Rutherford countered by declaring, "Bohr is different. He's a *football player!*" (Bohr's younger brother Harald was a goalie for the Danish Olympic team, and the young Niels was a talented player in his own right.)

Despite the huge difference in their basic temperaments, Bohr and Rutherford became great friends. And the young Dane was able to rescue Rutherford's solar-system atom, though only through desperate measures similar to those used by Max Planck to explain the black-body spectrum. Bohr recognized the atomic structure problem as a dramatic breakdown of classical physics akin to the black-body radiation problem. Just as in the "ultraviolet catastrophe," where classical physics says that hot objects should emit enormous amounts of short-wavelength light that they clearly do not, classical physics says that a nuclear atom can't exist for long, even though atoms are manifestly stable. Like Planck before him, Bohr came up with a new model for the atom by simply declaring that, under certain circumstances, the rules of classical physics do not apply.

The key to Bohr's model of the atom is the idea of "stationary states." Classical physics tells us that an orbiting electron should emit radiation, but Bohr suggested that for certain special orbits—similar to the "allowed modes" in Planck's black-body solution—an electron does not radiate. In the same way that Planck's imaginary oscillators could only emit energy in discrete multiples of a fundamental energy, Bohr's electrons can only move around the nucleus in orbits with discrete multiples of a fundamental angular momentum. Angular momentum is a quantity associated with a rotating object that takes into account both speed and the distribution of mass, and for an object that's not subject to significant outside forces, it remains constant. The classic example is a spinning figure skater: when skaters spin with arms extended, they rotate slowly, but when they pull their arms in, they spin faster. The angular momentum is the same in both cases, but as the distribution of mass changes, the rotation speed increases to compensate. For a particle in a circular orbit, the angular momentum is equal to the particle's linear momentum (mass times velocity) multiplied by the radius of the orbit, so for a given angular momentum, a particle could be orbiting slowly at large radius or rapidly at a small radius.

Bohr's "stationary states" were determined by a quantum condition similar to that used by Planck: an allowed orbit is one where the speed of the electron and the radius of the orbit are such that the angular momentum is an integer multiple of Planck's constant divided by 2π.[*]

Starting from this quantum condition, Bohr determined the properties of these stationary states by using classical rules to calculate the attractive force between the positive nucleus and negative electron, and the centripetal force needed to hold a particle in a circular orbit. While a particle orbiting quickly at a smaller radius might have the same angular momentum as one orbiting more slowly at a greater radius, it will require a much larger force to bend it around the smaller path—if you cut the radius in half, the speed will double, but keeping it in orbit will take eight times as much force. In a hydrogen atom, the force keeping the electron in orbit comes from the electromagnetic interaction between nucleus and electron, the behavior of which is well understood—cutting the radius in half increases the force by a factor of four. Putting those effects together gives a single optimum speed and radius for any particular value of angular momentum: once you use Bohr's quantum condition to pick a value of angular momentum, there is only one orbital radius for which the electromagnetic force is strong enough to hold an electron in orbit at the correct speed to produce that angular momentum.

These calculations predict a radius for a hydrogen atom[†] consistent with what was known in the early 1900s to be the approximate size of the atom. Knowing the speed of the electron lets you calculate its kinetic energy, which, combined with the electromagnetic attraction of the nucleus, tells you how much energy you would need to put into the atom to remove the electron entirely—how much extra kinetic energy the

---

[*]   This quantity recurs so frequently in the math of quantum physics that it gets its own symbol, $\hbar = h/2\pi \approx 1.055 \times 10^{-34}$ $J\text{-}s$ (joule-seconds), which physicists colloquially refer to as "h-bar."

[†]   This is now known as the "Bohr radius" in his honor, with a value of 0.0000000000529 meters. Atomic physicists regularly discuss distances involved in atomic and molecular interactions in terms of multiples of the Bohr radius.

electron would need to escape the attraction of the nucleus. The value Bohr calculated for this "ionization energy" matched the experimental value for hydrogen. Those results serve as a useful "sanity check," suggesting that the model is on the right track. The end result is a set of stationary states, each defined by an integer number of units of angular momentum, which ends up giving a well-defined energy for each state.

The energy of an electron in orbit around a nucleus is a combination of its kinetic energy due to its motion, and the potential energy due to its attraction to the nucleus. By convention in physics, the kinetic energy is always positive, while the potential energy is negative and depends on the separation between the electron and the nucleus. The electron's potential energy increases as it moves away from the nucleus, rising to nearly zero as the separation becomes huge, and diving toward negative infinity when the electron is right on top of the nucleus. This convention allows a clear distinction between states where the electron and nucleus are bound together to make an atom, and where the electron is merely passing by and has a chance to escape: if the sum of kinetic plus potential energy is negative, the electron will always be somewhere near the nucleus, and thus we say it's bound into the atom.

Bohr's quantum conditions, combined with the classical physics for a particle in orbit, give a set of orbits each having a negative total energy, following a simple pattern: the energy of the nth state is equal to the ionization energy divided by n^2:

$$E_n = -\frac{E_0}{n^2}$$

This corresponds to a set of circular orbits with increasing radii and energies that increase toward zero. There are also large ranges of energies that are simply impossible—an electron with one of those energies cannot satisfy Bohr's quantum condition.[*]

---

[*]   The separation between the energies of neighboring orbits decreases as the energy increases, so for very large n, they begin to blur together, but high-

Bohr's model describes orbits in which the electron is stable by fiat and does not emit any light. To get the spectrum of light emitted or absorbed by an atom, Bohr then applied the same rule used by Planck and Einstein to relate the frequency of light to an energy. In Bohr's model, light is emitted during quantum jumps from one orbit to another: when an atom emits light, an electron drops from a high-energy orbit to a lower-energy one, and when an atom absorbs light, an electron moves from a low-energy orbit to a high-energy one. (We'll discuss what triggers these jumps between states in Chapter 5.) In both cases, the change in the energy of the electron is accounted for by the energy of the light, which is related to the light's frequency by Planck's rule.

What determines the spectrum for hydrogen is not the energy of a given orbit, but the change in energy as the electron moves *between* orbits. The discrete orbits of the Bohr model lead directly to a discrete set of lines at particular energies in the spectrum, and give a simple explanation for the Rydberg formula, $1/\lambda = R \ (1/m^2 - 1/n^2)$: on the left-hand side of the equation, $1/\lambda$ relates to the energy of the emitted photon, while on the right, the one-over-integer-squared terms correspond to the energies of Bohr's stationary states. The constant R is just the ionization energy for hydrogen divided by Planck's constant and the speed of light, values that check out very nicely. The various series of spectral lines correspond to groups of transitions where electrons end up in a particular orbit, as illustrated in the figure below: the visible Balmer series involves atoms emitting a photon and ending up in the $n = 2$ state, while the ultraviolet Lyman series involves atoms ending up in $n = 1$.

Bohr's model also relates the constant R in Rydberg's formula to fundamental physics quantities like the electron mass and charge; this might not seem like that big a deal, but there are few things theoretical physicists like less than arbitrary new constants whose origins can't be traced to anything else. This allows Bohr's model to be extended to ions of heavier elements, with all but one of their electrons removed. According to the model, the energy of the stationary states should

---

precision spectroscopy has been used to study the properties of "Rydberg atoms" with n values running into the hundreds.

depend on the square of the charge of the nucleus; this insight was crucial for understanding the spectrum of x-rays emitted by different elements, and helped explain the organization of the periodic table, as we'll discuss more in Chapter 6.

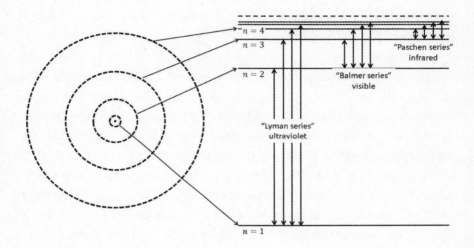

Orbits and energy levels in the Bohr model, with the transitions leading to three series of spectral lines labeled.

Of course, there's still one problem with the Bohr model, like the Planck model of black-body radiation that helped inspire it: there's no obvious reason to introduce the quantum hypothesis of stationary states. As long as you're willing to accept it, though, Bohr's model succeeds brilliantly for hydrogen and hydrogen-like ions; that may seem a modest success, but as it was the first success in decades, it started a revolution. Other physicists, notably Arnold Sommerfeld, found ways to formalize Bohr's quantum idea mathematically, and this quickly became the dominant framework for understanding the structure of atoms and molecules.*

---

* This is now known as the "old quantum theory." We'll talk more about the Bohr-Sommerfeld atom, and how it was replaced by modern quantum theory, in later chapters.

The greatest success of Bohr's model, though, was conceptual: it introduced the idea of discrete energy states within atoms, building on Planck's quantum hypothesis and Einstein's model of light quanta. While the mathematical techniques used to determine those atomic states and their energies have changed enormously, that central concept remains, and it's absolutely fundamental to our modern understanding of physics and chemistry.

Essentially everything we know about the structure of atoms and molecules comes from using the light they emit to deduce the energies of their allowed states; for heavier atoms, the spectrum can be very complicated, and provides a wealth of information about the arrangement of electrons and the interactions between them. Just as Planck's black-body spectrum allows us to determine the temperature of distant objects in the universe, the characteristic absorption and emission lines of various elements allow us to determine what those objects are made of. Here on Earth, too, many chemical analysis techniques rely on identifying the spectral lines of particular atoms and molecules.

These spectral lines also find technological applications in the course of an ordinary day—for instance, if your office has fluorescent lighting. Fluorescent bulbs contain a gas of mostly mercury atoms. When excited by an electrical current, these atoms emit light in the red, green, and blue regions of the spectrum, producing light that appears bluish-white to our eyes; they also emit a fair amount of invisible ultraviolet light, and fluorescent tubes are coated with a chemical that absorbs energy from ultraviolet light and emits it in the visible range, boosting the amount of light produced and—depending upon the specific coating—allowing lighting designers to control the mix of colors to produce different effects.

The high efficiency of fluorescent lights, too, is ultimately a function of Bohr's quantum condition. An incandescent bulb must heat its filament to a temperature high enough to generate a black-body spectrum with the desired color, but the emitted spectrum will necessarily include a large amount of infrared light that our eyes don't register. The gas in a fluorescent tube is diffuse enough that the atoms are essentially independent of one another, so they emit light not in a broad spectrum, but in discrete lines concentrated in the visible region. As a result, while the total amount of light generated for a given current may be less, a greater fraction of that light is visible to humans, so the overall efficiency is greater.

ATOMIC CLOCKS

Bohr's model of the atom, and the information it gave us about the spectrum of light emitted by atoms, also laid the foundation for a revolution in the measurement of time—it's why today even a cheap alarm clock can be traced back to quantum roots. The frequency of the light absorbed or emitted by an atom of a particular element is determined only by the difference in energy between two states of the electron, and those states are fixed by the laws of physics. Every cesium atom in the universe is identical to every other, and so they act as perfect little frequency references: if a cesium atom absorbs light, you know absolutely and without question what the frequency of that light is. Finally, then, we have a light source we can use as the basis for a clock.

The modern definition of the second is the time required for 9,192,631,770 oscillations of the light associated with a transition between two particular electron states in cesium.* A state-of-the-art, modern atomic clock consists of a microwave light source in a laboratory, and a collection of several million cesium atoms cooled to within a few millionths of a degree above absolute zero, which serve as the frequency reference. A cloud of these cold atoms prepared in one electron state is launched upward, passing through a cavity where the atoms interact with light from the microwave source. Then the atoms slow under the influence of gravity, and eventually fall back down through

---

* The states in question are not different orbits in the way Bohr originally envisioned, but "hyperfine" states with energy splitting dependent on the intrinsic "spin" of the electron, a property that wasn't discovered until 1922. The central concept remains the same, though: the frequency of the light depends on the energy difference between the two states in exactly the way Bohr described.

the cavity again. This second pass gives a second interaction with the microwaves, after which the atoms are measured to see what state they're in. If the frequency of the microwave source perfectly matches the frequency associated with the cesium transition, all of the atoms will have transitioned to the second state, while a small frequency error will result in some of the atoms being left behind in the initial state. The clock operators use the fraction of atoms making the transition to determine how to adjust the microwave frequency to better match the cesium transition, and the process repeats.

This two-interaction process (for which Norman Ramsey shared the 1989 Nobel Prize) is essentially the same as that you use to adjust a watch. First, you synchronize your watch with an accurate time reference—the National Institute of Standards and Technology official time service webpage, say. Then you wait a bit, and return to check your watch against the time reference; if it's running fast or slow, you adjust it to correct the time, then repeat the process.

In a cesium atomic clock, the first interaction with the microwaves plays the synchronization role, attempting to put the atoms into a state that oscillates at the exact frequency determined by the energy differ-ence between levels. The atoms and the microwaves start out exactly in phase, and then oscillate for some time before interacting again. If the frequencies match, the oscillations remain in phase, and all of the atoms end up in the second state; if the frequency is a little too high or low, some of the atoms remain in the initial state, and the physi-cists know that they need to adjust the frequency to compensate. Each cycle of the clock takes about one second, and after an hour or so of clock operation, the end result is a microwave source that matches the cesium transition frequency to within a few parts in 10^{16}. Such a "clock" would need to run continuously for billions of years before it would deviate from a clock based on the true cesium frequency by one second.

The official time for the world, defined by international treaty, is determined from a collection of more than seventy atomic clocks oper-ated by national laboratories in various countries. The name of this official time, UTC, is a fine example of international negotiation: in English, this would be called "Coordinated Universal Time," while in

French it would be "Temps Universel Coordonné." The final abbreviation is a compromise that doesn't make sense in either language. The official network time used to coordinate communications on the internet and other global communications networks is closely synchronized to UTC, so if you pull out your smartphone to check the time, it ultimately traces back to cesium clocks.

Of course, my cheap bedside alarm clock is not connected to the internet. It takes its time signal from the alternating electric current from the wall plug, which oscillates from high voltage to low and back sixty times per second. But even this can be traced back to atomic time—since modern power grids connect many power plants over huge ranges of space, the 60 Hz frequency of the power they provide is tightly regulated, and electric power companies rely heavily on atomic time and time distribution networks to help keep all their plants in sync. Without careful frequency control, a hydroelectric plant in Vermont might drift out of sync with one in Buffalo. Eventually, the company supplying power to my house in Niskayuna might find Buffalo trying to increase the voltage at the same instant that Vermont tried to decrease it. Those out-of-phase voltage oscillations would partially cancel each other out, reducing the total power available and leading to losses in the power grid that could cost millions of dollars.

In the end, then, all modern timekeeping—from the national labs monitoring chilly collections of cesium atoms to the networked computers time-stamping our email and even the seemingly low-tech alarm clock whose beeping starts my day—is fundamentally quantum. Like the builders of Newgrange, we mark the passage of time with light, but our clock operates on a much smaller and stranger scale: by counting the oscillations of light waves produced by electrons jumping between the atomic quantum states that Niels Bohr first proposed in 1913.

CHAPTER 5

THE INTERNET: A SOLUTION IN SEARCH OF A PROBLEM

*My **social media feeds** are full of the usual overnight fare—morning news from Europe and Africa, evening stories from Asia and Australia, digital photos of the kids and cats of friends who live around the world . . .*

No collection of technologies defines the current moment in history quite so definitively as the internet. The ability to communicate almost instantaneously with virtually anyone on the planet has radically changed not just communication itself, but any number of everyday activities that rely upon it. We buy music and movies, order just about anything to be delivered to our doorstep, and share messages and pictures with far-flung friends and family. Over an astoundingly short span of time, the internet has risen from something used by only a handful of researchers to an all-encompassing network affecting every aspect of life. We're still sorting out whether

75

the changes it has wrought will end up being a net positive, but the internet has clearly already transformed society, and will continue to do so for some time yet.

Long-distance telecommunication itself is not new technology—we've been sending electronic messages between continents since the days of the telegraph. The internet as we know it, though, would be impossible without the development of high-bandwidth fiber-optic networks capable of carrying enormous amounts of data. These days, most long-distance internet traffic is carried by pulses of light traveling down glass fibers, and the lasers that produce those pulses would be impossible without an understanding of quantum physics.

THE WEB BEFORE THE WEB

The era of global telecommunications is considerably older than many people realize, stretching back to 1858 and the completion of the first transatlantic telegraph cable between Ireland and Newfoundland. The initial connection required heroic efforts and lasted only about a month before it failed. For a brief moment, however, Europe and North America could exchange messages without waiting weeks for a ship to physically cross the ocean.

The brief success and early failure of the first cable spurred new efforts, and in 1866 a much more robust (and better engineered) cable was laid across the floor of the North Atlantic, and telegraphic contact between the continents has been maintained ever since. In the last century and a half, many more cables have been strung, connecting the entire globe.

The crucial metric for any communications network is the rate at which it can transmit information, often called its "bandwidth,"* which

---

* This is slightly confusing, as "bandwidth" is also used in communications to describe the range of frequencies that can be successfully transmitted through some channel.

is expressed in terms of bits per second.* The bandwidth of the initial transatlantic cable of 1858 was pretty terrible—the official first message, from Queen Victoria of Britain to President Buchanan in the US, took seventeen hours and forty minutes to send, well under a tenth of a bit per second. Improvements in cable engineering and telegraph technology rapidly increased these speeds—the 1866 cable already carried messages around eighty times faster than the 1858 one—but transatlantic bandwidth remained low well into the twentieth century.

Telegraph and, later, telephone cables carried electrical impulses over long strands of copper wire and faced severe problems with signal attenuation. Even an excellent conductor like copper has some electrical resistance, and over long distances this leads to a slow decrease in the voltage of the signal received relative to the voltage sent. This can be addressed by increasing the sending voltage, but only within limits—the ultimate failure of the 1858 cable was due in part to the unwise use of high-voltage sources on the North American end, which eventually compromised the insulation on the underwater cable.

While signal attenuation was a problem for cables on land as well, it was especially challenging for those stretching under oceans. On land, attenuation can be addressed by adding "repeaters" at regular intervals, to receive a low-voltage signal and retransmit it with a higher voltage. Placing repeaters in the middle of hundreds of kilometers of ocean was completely impossible in the 1860s, however, and it was nearly a century before the first cable with automatic repeaters was strung across the Atlantic. And though adding repeaters does address the attenuation problem, it adds to the cost and complexity of cables both on land and under the sea. Finding clever ways to boost the bandwidth of copper transmission lines remained a major problem for telecommunications engineers for many decades.

---

* Older sources often use "words per second," but this is a little ambiguous because, of course, words vary enormously in length. The modern method of quantifying information in terms of 0 or 1 binary bits was developed by Claude Shannon in the 1940s and is much more reliable.

The development of lasers allowed a dramatic increase in bandwidth by changing to an entirely different method of signal transmission. Rather than encoding the "0" and "1" of a signal's bits as different voltages sent via copper cable, modern networks represent them with on-or-off pulses of light sent down thin fibers of glass.

An optical fiber consists of a thin cylinder made of two slightly different types of glass, with a thin "core" of one type surrounded by a "cladding" of the other. Light traveling through the core can reflect off the boundary between the two, effectively confining it to the core, even as the fiber bends around corners. This allows light pulses to be steered along arbitrary paths, without needing a straight line of sight from one end to the other.

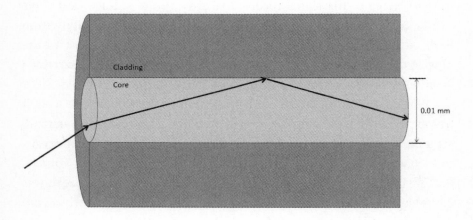

The structure of an optical fiber, with a glass core surrounded by cladding of a different type of glass. Light rays entering the end of the fiber reflect off the boundary and are confined to the core.

Optical fibers offer a huge advantage over copper wires in terms of signal attenuation. Light pulses sent through fibers do decay, as some light leaks out or is absorbed in the glass, but at the infrared wavelengths used in modern fiber systems—the two primary bands use light at around 1,300 nm or 1,500 nm—the distance a signal can be sent before needing a repeater is more than ten times greater than the distance for copper transmission lines. Optical fibers can also be packed together much more densely than copper wires, as the light confined to the core of one fiber has no way to reach the core of another fiber nearby. This eliminates the problem of cross talk between neighboring conductors, where a high-voltage signal in one copper wire can induce a smaller signal in another wire if the two are too close together.

The shift from sending signals as electrical pulses through copper wires to light pulses through glass fibers has led to an explosion in the bandwidth available for global telecommunication networks. I'm (barely) old enough to remember a time when it could be difficult to make voice calls between countries, while my kids take for granted the ability to watch live high-definition streaming video from almost anywhere.

For fiber-optic networks to be viable, though, required a dramatic leap in the technology used to generate and manipulate light. In particular, high-bandwidth fiber optics demand a source that produces a single frequency of light in a beam that can be steered into a fiber core thinner than a human hair. No classical source of light will do: a hot object produces light over too wide a range of frequencies to be useful, and while the spectral lines from a gas of atoms—as discussed in the previous chapter—might be narrow enough, the light of a gas (like that used in fluorescent tubes) is too diffuse to be efficiently coupled into the fiber.

The kind of light needed for high-bandwidth fiber-optic telecommunications can only come from a laser. And making a laser requires a detailed understanding of the quantum rules that govern how atoms emit light, an understanding that can be traced to a familiar source.

HOW ATOMS EMIT LIGHT

The first working laser was made in 1960, following competing theoretical proposals that led to a protracted patent fight. The essential physics, though, had been worked out over forty years earlier in a 1917 paper by Albert Einstein.

Einstein's primary claim to fame, both within the field of physics and among the general public, comes from his development of relativity, particularly general relativity, with its explanation of gravity as the four-dimensional warping of space-time by matter. This sometimes leads people to assume that he must have always worked with highly abstract mathematics, but this isn't true. His initial papers on special relativity in 1905 use relatively simple math, and it was another ten years before he completed general relativity in 1915. The decade-long gap between theories occurred in large part because he was laboriously learning the mathematics of curved spaces he needed to complete general relativity, with the assistance of his friend Marcel Grossmann. Einstein was highly capable mathematically, but his real genius lay in his intuition for physics and a certain clarity of insight. In the end, he was almost scooped on general relativity by the mathematician David Hilbert, who understood the necessary math far more thoroughly than Einstein did. Hilbert later said that "Every boy in the streets of Göttingen understands more about four-dimensional geometry than Einstein," but he did credit Einstein with having the physical insight that made the theory possible.

Einstein's formal background in physics was more in the area of what today would be called statistical mechanics: studying the properties of large collections of particles. His 1905 PhD thesis, "A New Determination of Molecular Dimensions" is surprisingly mundane—relating the viscosity of a sugar solution to the size of the dissolved molecules—compared to his most famous work. That same year, he followed up his thesis with a paper on Brownian motion, a kind of jittering motion observed in microscopic particles floating in water. Einstein attributed Brownian motion to random collisions between these particles and the surrounding water molecules, and using an equation he had derived in his thesis, showed how to use statistical measurements of Brownian motion to determine the properties of those molecules. Together, these two papers were highly influential, helping to convince the last holdouts that atoms and molecules were real, physical entities and not merely a calculational convenience.

Einstein's 1917 paper drew on these statistical roots to consider what happens when huge numbers of photons and atoms interact with each other. This might seem a quixotic project to take on, especially as it was still the era of the "old quantum theory," and neither photons nor atoms were fully understood. A common quirk of physics, though, is that problems that prove impossible when dealing with a few particles prove to be surprisingly simple when applied to an uncountably large number. The physics of a single photon interacting with a single atom may have been poorly understood, but thinking about vast numbers of them allows you to gloss over many of the details of the individual interactions. The same sort of statistical reasoning that let Einstein connect Brownian motion to molecular properties without knowing the details of any individual molecular collision allowed him to deduce some properties of photons from an extraordinarily basic model of their interactions with matter.

Einstein's paper considered the interaction between photons and a collection of Bohr-type atoms that absorb or emit light only when an electron moves between two discrete allowed orbits. For simplicity, he considered only two states of the atom (a low-energy "ground state" and a higher-energy "excited state"), so he only had to keep track of a single frequency of light, determined by the energy difference between the two states.

In this simple picture, interactions between the light and the atom can be classified in terms of two conditions: whether the atom is in the ground state or the excited state, and whether there is light of the appropriate frequency present. Within this scheme, there are three possible processes that can occur:*

1. **Absorption**—If you have an atom in the ground state and light of the appropriate frequency, an atom can absorb a photon and move to the excited state.
2. **Spontaneous Emission**—If you have an atom in the excited state, it can drop down to the ground state and emit a photon, whether there is already light present or not.

---

* The fourth possible configuration would involve a low-energy atom spontaneously changing states in the absence of light, which is impossible because it would violate the law of energy conservation.

3. Stimulated Emission—If you have an atom in the excited state, a photon of the appropriate frequency can trigger it to emit a second photon and drop to the ground state.

The first two of these were already well-known in 1917,[*] as absorption and emission of light by atomic vapors were being used to identify elements long before Bohr's quantum model of the atom. The third process, stimulated emission, was Einstein's own invention, and turns out to be the critical piece of physics that makes the laser (and thus the modern internet) possible.

It may seem strange to think of one photon causing the emission of another, sending energy in and somehow *lowering* the energy of the atom, but as Einstein pointed out, if you think of the electron in an atom as an oscillator (which it must be, in some sense, to generate light), classical physics demands that this process should exist. It's easy to picture by using the analogy of pushing a child on a swing: if you time your pushes to come as they reach the highest point of their swing, you'll increase the energy in their swinging motion and drive them higher. If, however, you push them with exactly the same frequency, but with the pushes timed to be against their motion as they pass through the lowest point of the swing, you'll quickly bring them to a stop.[†] In the same way, light of the right frequency "pushing" on an orbiting electron ought to be able to either increase or decrease the energy of the electron. In the quantum scenario, a decrease in energy from the upper to the lower state must lead to the emission of a photon.

---

[*]   While the idea that atoms should spontaneously emit light and drop to lower states was well established, it turns out to be far and away the hardest of these processes to explain mathematically. A full understanding of why spontaneous emission happens requires the complete theory of quantum mechanics; loosely speaking, the emission is triggered by energy present in empty space. This wasn't worked out for another couple of decades, though—Einstein merely took the spontaneous emission of light from excited atoms as an empirical fact.

[†]   This is much less popular with actual children on actual swings than as a thought experiment.

While Einstein was not able to spell out all the details of stimulated emission, the classical analogy tells us that stimulated emission should act to amplify the light that's already present: the emitted photon must have the same frequency as the photon that triggered the emission, and it must be moving in exactly the same direction. In short, stimulated emission is a process that takes one excited atom and one photon and produces one ground-state atom and two photons that are identical in every way.

WHAT EINSTEIN LEARNED ABOUT LIGHT

Having identified these three processes, Einstein skipped past the details of how they might work, and simply declared that each must happen with some probability. He then drew on his background in thermal and statistical physics to see what he could deduce about those probabilities—and the properties of photons—from features that might be observed in an extremely large collection of atoms interacting with light. Considering this simple model of atom-photon interaction in terms of probabilities, Einstein uncovered a wealth of physics.

The critical principle was borrowed from thermodynamics, namely the idea that a gas of atoms and a collection of photons ought to be able to reach a state of equilibrium. In equilibrium, the overall properties of a large system are not changing, even though the individual components may be—when two atoms in a gas collide, if one atom slows down, the other speeds up, so the total energy of the gas (and thus its temperature) remains constant. Equilibrium states are the foundation of thermodynamics and statistical physics, and a powerful tool for reasoning about the properties of large collections of atoms and molecules; it was only natural for Einstein to extend this idea to include light quanta as well.

In the simplified atom-photon model Einstein used, equilibrium would mean that any photons absorbed by one atom will be shortly replaced by photons of the same frequency emitted by some other atom, and any atom that drops from the high-energy state to the low-energy one will be soon replaced by a new atom excited to the high-energy state by absorbing a photon. In such a state, both the number of high-energy atoms and the intensity of the light remain constant, on average. The question, then, is what properties light must have in order for a gas of atoms starting at some temperature to reach equilibrium with that light.

In ordinary thermodynamics, we generally find that equilibrium occurs when the different components of a system reach the same temperature. If you place a piece of hot metal in cold water, for example, the state of the system will change very rapidly at first, with the metal cooling and the water heating. Once both metal and water reach the same lukewarm temperature, though, they will stop changing, having reached equilibrium. One of the questions Einstein considered was whether the same would hold true for a mixture of atoms and light.

We've already seen one way of associating a temperature with light: Planck's description of black-body radiation, whose spectrum is determined only by the temperature. We can also assign a temperature to the atoms in two ways: the first is the familiar definition of the average kinetic energy of atoms moving in the gas, but the temperature is also reflected in the number of excited-state atoms present. Some of the thermal energy of the gas can be converted to internal energy of the atoms, for example via collisions between two ground-state atoms that leave both atoms moving more slowly, with one now in the excited state. For a gas of atoms with a given temperature, the probability of finding any specific atom in the excited state is a simple function of the temperature, which was worked out by Maxwell and Boltzmann in the late 1800s.

Starting with a gas of atoms at some temperature, interacting with light via the three photon processes above, Einstein showed that the number of photons present—that is, the intensity of the light at the relevant wavelength—when the system reaches equilibrium exactly matches the predictions from Planck's formula for the spectrum of a black body at the same temperature as the atoms. Similarly, if you start with a black-body spectrum for the light, and all the atoms in the lowest-energy state, at equilibrium the number of atoms in the upper state is exactly what you would expect to find in a gas at the appropriate temperature.

The fact that Planck's quantum formula for the black-body spectrum emerges naturally from applying the quantum idea to light was a powerful argument in favor of the reality of photons. Of course, for a gas of atoms to reach equilibrium with light also demands that absorbing and emitting photons can change the velocity (in order to change the average kinetic energy) of the atoms. This in turn implies that individual photons must carry momentum, and Einstein used his model to show that the necessary photon momentum exactly matches what you would expect from his 1905 theory of special relativity, demonstrating that quantum light is consistent with yet another well-established field of physics and providing additional support for the notion of photons.

The momentum of a photon was directly observed a few years later, by Arthur Holly Compton, as a change in the wavelength of x-rays that bounce off the electrons in a metal. The experimental observation of this "Compton scattering" was one of the final pieces of evidence sealing the case for the particle nature of light,* and won Compton a share of the 1927 Nobel Prize in Physics. These days, photon momentum is an essential tool for the technique of laser cooling, which uses light scattering to slow the motion of atoms in a gas, producing small clouds of atoms at temperatures within a millionth of a degree of absolute zero. These techniques have revolutionized the study of atomic and molecular physics, as the properties of such slow-moving atoms can be measured with unprecedented precision, and the 1997 Nobel Prize in Physics went to three physicists† for developing laser-cooling techniques in the early 1980s.

---

* Later work showed that it's possible to explain the Compton effect with a semiclassical model of light as a wave, but the photon picture is much simpler. The existence of photons as real particles was only incontrovertibly shown in a 1977 experiment that has no classical analogue.

† The 1997 Nobel Prize winners in Physics were Steve Chu of Stanford (later US Secretary of Energy in the Obama administration), Claude Cohen-Tannoudji of the École Normale Supérieure in Paris, and my PhD thesis advisor, Bill Phillips of the National Institute of Standards and Technology in Maryland.

Einstein also used his statistical model to show a simple and direct relationship between the rates of spontaneous emission, stimulated emission, and absorption. In order for the mixture of light and atoms to come to equilibrium, the rates for stimulated emission and absorption must be equal to one another, and proportional to the rate of spontaneous emission. An atom with a high rate of spontaneous emission will also readily absorb light, and an atom that readily absorbs light can easily be stimulated to emit light.

The exact rate of spontaneous emission for a particular atom was impossible to calculate in 1917, and would need to wait at least a decade for the development of the full theory of quantum mechanics. The relationship that Einstein found between how readily atoms absorb light and the spontaneous emission rate (generally measured in terms of the lifetime of an atom excited to a particular state) is empirically testable, though, and holds up very well. The model also predicts that the spontaneous emission rate should increase rapidly with the frequency of the emitted light, and in fact experimental observations have found this to be the case.[*]

Einstein's 1917 paper on the statistics of light is not his most famous work, but it was an essential piece of the foundation for the field of quantum optics. The simple probabilistic model of absorption, stimulated emission, and spontaneous emission is still used to predict interactions between light and a gas of atoms, and the labels for these probabilities are called the "Einstein coefficients" in honor of this paper. Perhaps most significantly for physics as a whole, the paper played a crucial role in convincing physicists to take photons seriously, at a time when even Niels Bohr was reluctant to accept them and preferred a more classical model in which his discrete atomic states interacted with light that was only a wave.

---

[*] The correlation isn't perfect, as there are other effects that can lead to very long lifetimes even for states that emit visible light. Again, a complete explanation of the lifetimes of particular states only became possible after the completion of quantum mechanics.

For our purposes, though, the most important part of Einstein's 1917 work on photons is in the setup: the introduction of stimulated emission. The fact that one photon can trigger the emission of a second photon just like it is what makes the laser possible, with dramatic consequences for everyday life.

LASER HISTORY

Like a lot of physicists, Charles Townes spent World War II working on the new technology of radar, which led to dramatic improvements in methods of generating, controlling, and detecting light at frequencies in the microwave region of the spectrum. After the war, physicists returning to peaceful research began to use these new microwave sources to investigate the properties of atoms and molecules, mapping out transitions between states. These investigations led to revolutionary developments in physics, such as when Willis Lamb and Robert Retherford discovered a small energy difference between two states in hydrogen that should have been identical. Trying to explain this "Lamb shift" led to the development of quantum electrodynamics (QED), one of the strangest theories in science, but also arguably the most precisely tested in history.*

Microwave spectroscopy experiments were also the first step toward the development of the laser, as Townes and others searched for ways to extend the range of light wavelengths they could study to lower frequencies (longer wavelengths) than those used in wartime radar development. Lower frequencies were of interest because many molecules absorb and emit light in this part of the spectrum, and Townes hit on the idea of using the molecules themselves to generate the microwaves.

---

*   QED is a little too exotic to have many everyday consequences, so we won't talk about it in detail here. You can learn a bit more in *How to Teach Quantum Physics to Your Dog*, or a lot more from Richard Feynman's *QED: The Strange Theory of Light and Matter*.

Townes produced a beam of ammonia molecules in an excited energy state, and sent them through a microwave cavity—a metal chamber with a small hole, like the imaginary box we used to set up Planck's black-body model in Chapter 2. The size of the cavity was chosen to correspond to the wavelength of the microwaves emitted by ammonia molecules. Any photons that happened to be emitted by the ammonia molecules passing through would happily bounce back and forth inside the cavity, remaining there for a long time before escaping through the hole.

By itself, this would not be terribly interesting, as the rate of spontaneous photon emission by molecules at that wavelength is rather small. But thanks to the process of stimulated emission, their device acted as an amplifier. An excited ammonia molecule entering the cavity could encounter a photon already inside, at exactly the right frequency to (potentially) stimulate the emission of a second photon identical to the first. Subsequent molecules would find two photons inside, making stimulated emission even more likely, and as the process repeated over and over, the number of photons would grow. Townes described this with an acronym: MASER, for Microwave Amplification by Stimulated Emission of Radiation.

Townes's maser produced a relatively intense source of microwaves in an extremely narrow range of frequencies, a possibility that follows logically from Einstein's 1917 photon model, though Einstein had not considered it in his paper. In an ordinary gas, the vast majority of the atoms are in low-energy states, so it's relatively rare for a photon to encounter an excited atom and cause stimulated emission. In his maser, however, Townes used a stream of molecules that he'd already excited with an electrical current, meaning they were mostly in the *higher* energy state, an unusual arrangement termed a "population inversion." This inversion makes any photons in the cavity more likely to encounter an excited molecule and stimulate it to emit. Each new photon is identical in frequency (and direction of motion, polarization, and other optical properties) to the one that stimulated it. And because each of those photons can in turn stimulate the emission of another identical photon, the process leads to an exponential growth in the number of photons (one begets two, beget four, beget eight, etc.) within a very narrow range of wavelengths.* A tiny fraction of the light that builds up can be extracted through small holes in the cavity, and its frequency measured with high precision—masers using hydrogen atoms are a crucial element of the system used to determine and disseminate time from atomic clocks, helping to keep time between cycles of the cesium clock.

Following the development of the maser, Townes began discussing how to extend the basic idea to visible regions of the spectrum with his colleague (and brother-in-law) Arthur Schawlow, among others. Townes and Schawlow would eventually hit on the trick to making an "optical maser," though they were beaten to the idea by a graduate student named Gordon Gould, who gave the device its modern name. After a conversation with Townes, Gould wrote down some ideas in a notebook,† under the heading "LASER: Light Amplification by Stimulated Emission of Radiation," a name that has stuck (though few remember its origin as an acronym), and eclipsed the original "maser."

The key components of a laser are the same as for the maser: a "population inversion" with a lot of electrons in higher-energy states within atoms or molecules associated with the frequency of interest, and a cavity to keep the emitted photons bouncing around and interacting with those atoms. As we'll see, obtaining these components is slightly more complicated for visible light than microwaves, but once they're in place, the mechanism is the same: photons already in the

---

*   It's important to note that this is not an equilibrium situation. Maintaining the population inversion requires a constant input of energy from some other source (in Townes's original ammonia maser, this was in the source of the molecular beam); without the inversion the maser operation will quickly cease, and the system will settle down to an equilibrium distribution with mostly low-energy atoms and a black-body radiation field at some moderate temperature.

†   This notebook page, which Gould had notarized, went on to play a crucial role in the court cases that ultimately secured patents for Gould on several essential laser technologies. In the interests of full disclosure, I should note that Gould was an alumnus of Union College, where I teach, and the Department of Physics and Astronomy has an endowed chair in his honor.

cavity trigger stimulated emission from the excited atoms, and the number of photons grows exponentially.

The first technical obstacle to making the move from maser to laser was generating the population inversion. Most excited states with energies corresponding to the frequency range of visible light have extremely short lifetimes before they spontaneously emit a photon (as predicted by Einstein's model) and move to a lower-energy state, making it difficult to keep excited atoms around to be stimulated. States with long lifetimes that would more easily sustain an inversion are difficult to excite directly (again, as predicted by Einstein's theory). This problem is generally solved by using a multilevel scheme in which electrons are excited by indirect means. For example, a helium-neon laser uses excited helium atoms to transfer energy to neon atoms as they collide in a plasma. This indirect process produces a population inversion with many more neon atoms with a relatively long lifetime in a particular high-energy state than a plasma of only neon would generate directly. The mix of helium and neon together in a plasma provides the gain medium for laser operation at the red wavelength familiar from early supermarket scanners.* As long as the current creating the plasma is maintained, helium atoms will continue to be excited and to excite neon atoms in turn, allowing continuous operation of the laser.

The other major technical obstacle to moving from maser to laser, and the sticking point for Townes, was constructing the cavity to catch the photons. A microwave cavity consists of a (nearly) fully enclosed space surrounded by metal walls with dimensions comparable to the wavelength of the microwaves themselves—a few centimeters for the masers Townes was used to working with—and only small holes to allow the introduction of excited molecules and the extraction of light. This concept doesn't carry over very well to optical wavelengths—even today, making a fully enclosed cavity with dimensions of only a few hundred nanometers would be very challenging, and in 1957 it would have been simply impossible.

---

* Modern systems use semiconductor diode lasers, which are much more compact than helium-neon lasers and work at very similar wavelength.

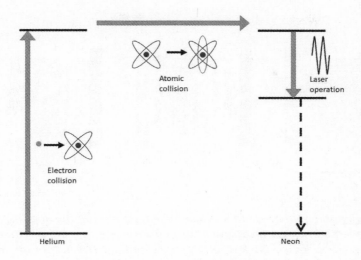

Operating scheme for a helium-neon laser. Helium atoms are excited to a high-energy state by collisions with electrons in a plasma. Collisions between helium and neon atoms excite the neon atoms to a long-lived state, creating a population inversion that's used to make a red laser.

The insight that allowed the invention of a working laser (realized by Gould and Schawlow, and also Aleksandr Prokhorov in the USSR) was that the cavity need not be so fully enclosed after all; it's sufficient to use two mirrors facing one another, which will bounce photons back and forth in a line between them. This more open structure allows plenty of room to contain large numbers of atoms or molecules—some gas laser systems use cavities a couple of meters in length—and also gives the laser one of its defining characteristics: because the cavity only traps photons along a single line, the light produced by a laser emerges in a single, narrow beam. (The beam emitted is a tiny fraction of the light within the cavity that escapes because one of the mirrors is just short of perfectly reflective, allowing a few percent of the photons hitting it to pass through.)

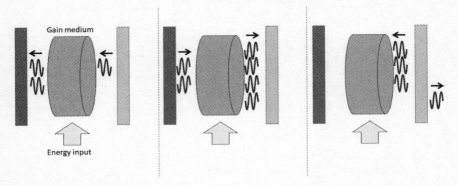

The cavity and gain medium in a laser. A single photon moving from right to left stimulates a second photon from the gain medium, then two photons are reflected back through to create four. The output beam of the laser comes from a small fraction of photons that leak out through one of the mirrors.

"A SOLUTION LOOKING FOR A PROBLEM"

With the idea of an open cavity in place, development of practical lasers moved ahead, with the first working laser built by Theodore Maiman at Bell Labs in 1960, using chromium atoms inside a rod of synthetic ruby as the amplifying medium. This first laser used xenon flashlamps to produce the population inversion: a sudden bright pulse of white light would excite the chromium atoms to high energy states, some of them falling into an energy state with a lifetime of about 5 ms (which is long by the standards of atomic physics). This produces a short-lived population inversion, leading to a brief pulse of laser light.

Over the next few years, numerous other laser types were developed, with gain media ranging from gases like the helium-neon laser described above, to liquids containing organic dye molecules (first demonstrated in 1966), to solid-state lasers using semiconductor materials (the first gallium arsenide laser was demonstrated in 1962). Semiconductor-based lasers have proven to be particularly important, as they're extremely compact—the size of computer chips—and built into all manner of consumer electronics. If you have a CD, DVD, or Blu-ray player, or even a laser pointer you use to annoy your pets, you're making regular use of semiconductor lasers.

In the very early days of laser physics, the devices were regarded as a bit of a curiosity, without much practical application—one of Maiman's assistants, Irnee D'Haenens, famously called the laser "a solution looking for a problem." This search did not take long, though, and over the last fifty-odd years, innumerable problems have lined up to be solved by lasers.

Within physics, lasers are invaluable tools for precision measurements. Because the photons in a laser are produced by stimulated emission, they are identical to a degree that can't be achieved using light from a lamp. Some laser sources can be tuned over some frequency range; spectroscopic measurements made with these lasers can pin down the precise characteristic frequencies of light absorbed and emitted by atoms to eighteen decimal places. The photons from a laser are also all emitted with the same phase—in wave terms, the peaks and valleys of the light waves all align—allowing laser-based sensors to measure changes of position within a tiny fraction of a wavelength. The ultimate example of precision-position sensing with lasers is the Laser Interferometer Gravitational-Wave Observatory (LIGO), which in 2015 used two huge detectors to measure the tiny stretching and compression of space-time caused by passing gravitational waves created in the collision of two black holes. The change in the distance between the mirrors caused by these waves was smaller than the width of a single proton, but was clearly detected by LIGO, making headlines around the world.

Outside of precision physics experiments, most commercial laser applications don't directly use the frequency and phase characteristics of the laser—they just require a bright source of light. The narrow phase and frequency spread produced by stimulated emission are essential even for these, however, because they allow the formation of an exceptionally narrow beam of light. While laser beams do expand as they travel, their width increases as slowly as is possible. The Apollo missions left suitcase-sized arrays of retroreflectors on the lunar surface, and for more than forty years scientists have been shooting lasers at these and measuring the round-trip time to determine the distance to the moon, which is increasing by about 3.8 centimeters per year. The laser beam expands from an initial diameter of about 3.5 meters to about 15 kilometers, but it takes a 770,000-kilometer round-trip to do that, so it's not surprising that the beam of a laser pointer shot across the room to tease a pet doesn't appear to expand at all.

Narrow laser beams are used in construction and surveying to provide straight and level lines across moderate distances, greatly simplifying the process of building structures with level floors. Pulsed lasers can also be used to measure distance by determining the round-trip time for a pulse to travel out to an object and reflect back; the same basic technique can also measure the speed of moving objects, to the chagrin of many a heavy-footed driver.

The narrow beams formed by laser light are also crucial for techniques for precision cutting of wooden and metal parts. A relatively modest amount of electrical current will power a laser that can produce a tiny spot of light intense enough to burn most materials. The laser can be steered and adjusted with lenses and mirrors, allowing precise control of its position, and as a laser beam lacks physical cutting surfaces, it's not subject to wear and produces more uniform cuts. Lasers are also used to cut human tissues in some medical procedures—most commonly eye surgeries, but increasingly in other fields as well. The extremely localized high temperatures involved in laser cutting cauterizes tissues as they are cut, which can significantly reduce bleeding.

Though the above applications represent only a small sample of the many problems solved by lasers, even these would be enough to make lasers a major and important technology. The most important use of lasers in today's world, however, is as the backbone of modern telecommunications, including the internet.

WEB OF LIGHT

Early in the chapter, we discussed the enormous boon fiber-optic networks have been to telecommunications. Light pulses sent down glass fibers have dramatically lower attenuation rates than electrical pulses along copper wires, allowing more reliable and higher-bandwidth communications over long distance, and modern fiber-optic technology would be impossible without lasers. The narrowness of a laser beam is essential, as typical optical fibers are about the thickness of a human hair with a core a tenth that size. Coupling even a laser into a core that small is no trivial matter,* and such a system would be impossible with any nonlaser light source.

The narrow wavelength and frequency spread of laser light gives fiber-optic telecommunications a further advantage when it comes to boosting bandwidth. As mentioned before, multiple optical fibers can be bound together without the problem of "cross talk" or signal leakage, as between copper wires in close proximity. What's more, even a single fiber can carry several different signals at once by encoding them with different lasers having very slightly different wavelengths. The laser beams can be combined before entering the fiber and separated at the receiving end, allowing a single strand of fiber to transmit something like twenty signals at once, vastly increasing the carrying capacity of telecommunications networks.

While the earliest computer networks were carried over copper transmission lines, the modern internet of streaming video and endless cute-cat photos on social media would be unthinkable without the explosion in bandwidth that followed the introduction of fiber-optical telecommunications in the 1980s. The first transatlantic fiber-optic cable, in 1987, could carry 40,000 simultaneous telephone calls, ten times the number as the copper cables that immediately preceded it. The most recent transatlantic fiber-optic link, completed in 2017, carries digital data at a rate of 160 trillion bits per second, more than

---

* In my research lab, we sometimes need to pass lasers through optical fibers to get them from one place to another, and it's not uncommon to take several hours to align the lasers well enough to get even half of the light through—it's a painstaking process. Modular telecommunications systems make this much easier, of course, and some telecommunications systems use lasers that are built directly into special optical fibers, but there are decades of engineering effort behind those systems.

500,000 times that of the 1987 cable, and more than a *quadrillion* times the rate of the first transatlantic telegraph message back in 1858.

The amount of data transmitted per month on the global internet of 2016—two years ago, as of this writing—was around 1,000 times greater than that transmitted in the entire year of 2000, and nearly all of the long-distance connections making up that network are laser pulses carried along optical fibers. So the next time you log on and admire the baby pictures sent by a friend on another continent, remember that you ultimately have Einstein, statistics, and the quantum nature of light and atoms to thank for it.

CHAPTER 6

THE SENSE OF SMELL: CHEMISTRY BY EXCLUSION

*My tea is still a bit too hot to drink, but I **savor the aroma** of the rising steam as it cools . . .*

When it comes to the detection of odors, humans are no great shakes, especially compared to our friends in the animal kingdom, who devote an amazing percentage of their brains to the processing of smell. While our noses are kind of underwhelming, though, smell still has a powerful hold on us, particularly when it comes to food. The odor of cooked food is an essential part of the experience of eating, and the absence of odor cues can change the apparent taste of substances. Taste-testing different vegetables with your nose plugged is a science-fair staple, and it's surprisingly difficult to tell the difference between, say, an apple and a potato if you can't smell them.

The detection of smells is a complicated chemical process, whereby smallish molecules wafting up from an object reach and trigger receptors in the nose. Digging into the details very quickly runs into a blizzard of intimidating chemical names ("2-ethyl-3,5-dimethylpyrazine" is one of the molecules responsible for the aroma of coffee, for example) and competing models of the exact mechanism by which odor receptors in the nose produce the sensation of smell we experience. It's a ferociously complicated subject, and the science is still not completely settled.

One thing we can say for sure is that at the deepest level, the process of detecting smell is inherently quantum. The chemistry of the molecules involved in smell—and indeed all of chemistry as we know it—is rooted in some of the strangest of quantum phenomena, in particular the bizarre property known as "spin."

HOW SMELLS WORK

The human sense of smell (and that of most other animals) works in a manner similar to the color vision system discussed back in Chapter 3. Small molecules entering our nostrils form chemical bonds with special "odorant receptor" molecules in the nose, which are connected to individual neurons high in the nasal cavity. When an odorant receptor bonds to a molecule from the air, it triggers the neuron to send a signal to the brain, which collects signals from all the different neurons, and then processes them into what we perceive as the smell of whatever is under our nose.

The perception of smell is vastly more complex than the perception of color, however. While the human retina contains only three different types of color-sensing cells, each sensitive to a fairly broad range of wavelengths, the human nose contains a few *hundred* types of odorant receptor neurons.* A given odorant molecule making it into the nose can trigger several receptors at once, and different combinations of receptors are registered as different smells. A molecule from

---

* This is a lot, but actually relatively small compared to our fellow mammals. Some species have up to 1,000 distinct types of odor receptors in their noses.

my favorite tea will trigger one set of receptors—#3, #17, and #122, say—while a molecule wafting from the coffee of the person next to me triggers another—#3, #24, #122, and #157. Some of the same neurons are involved, but the resulting combinations, and thus the perceived scents, are very different.

The larger variety of receptors leads to a vast number of possible scents that can be perceived. While studies of color vision suggest humans can distinguish several million subtly different colors, recent estimates put the number of odor combinations we can detect as high as a *trillion*.

Another difference between smell and vision is that the mechanism by which odorant receptors are triggered remains the subject of some debate. Color vision is well understood as a photon absorption process, in which a particle of light triggers a transition between states in a molecule that then kicks off the signaling of a neuron. Each of the individual photons being detected is completely described by a single frequency, making the response of the light-sensitive cells in the eye unambiguous and easy to predict.

Smell, on the other hand, relies on a chemical process to detect molecules that vary internally to a staggering degree, in sometimes subtle ways. Two molecules with similar composition in terms of the number of atoms of particular elements, can differ in the *arrangement* of those atoms, and those structural differences can lead to dramatically different properties. If you take one oxygen atom, two carbon atoms, and six hydrogen atoms, and put them together with the two carbon atoms off to one side of the oxygen atom, you get ethanol, which is a liquid at room temperature and the active ingredient in alcoholic beverages. If you take the same collection of atoms, though, and sandwich the oxygen between the two carbon atoms, you get dimethyl ether, which is a gas at room temperature and is used as a propellant in aerosol sprays.

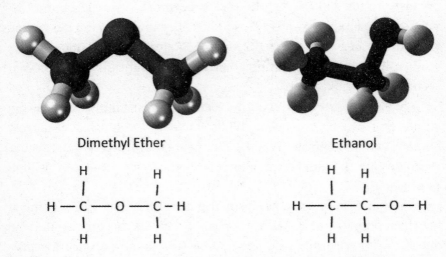

Dimethyl Ether   Ethanol

The three-dimensional structures of dimethyl ether and ethanol, two very different molecules with the exact same chemical formula.

The odor-detection system in our nose picks up on some of these subtle variations in the arrangement of atoms, letting us perceive chemically similar molecules as having very different smells. Historically, there are two competing theories about how the receptor molecules in the nose distinguish between different molecules. The more popular of the two is the "shape theory," which holds that different types of receptor molecules respond to the three-dimensional arrangement of atoms in the molecule being detected. The competing "vibration model" holds that receptor molecules distinguish their target molecules by the way they move—the atoms within a given molecule will shake back and forth at frequencies that are characteristic of that particular molecule and how its atoms are arranged. Proponents of the vibration theory argue that a given receptor is triggered by the presence of atoms vibrating within a particular range of frequencies.

Neither of these models is completely successful—each works well to explain the results of some experiments, and comes up short when applied to others. A full explanation of odor detection may turn out to involve elements of both, with some receptors primarily sensing shape while others pick up vibration.

Each model, though, is thoroughly quantum. Both the vibration frequencies of a particular molecule and the shape of the molecule itself depend on the three-dimensional structure of that molecule, which is determined by the quantum behavior of electrons that governs exactly how atoms bind together—how many other atoms a given atom can bond to, how strong those bonds are, what the angle between bonds will be, and so on.

Explaining those bonds requires a deeper look at the behavior of electrons than we can get from the Bohr model alone. It also requires the introduction of an entirely new property, one that has no analogue in classical physics, that will turn out to be essential for a wide range of everyday phenomena. To set the stage, though, we first have to take a brief excursion into the histories of chemistry and the classification of atoms.

THE PERIODIC TABLE

The periodic table of elements is a familiar sight—a mostly rectangular array of boxes with two small towers rising up on either end—and one of the most reliable visual indicators of a science classroom. The idea of a "Periodic Table of _____" is also a recurring visual joke online, with various sites offering "periodic tables" of basically anything that comes in multiple varieties.*

The periodic table as we know it is largely credited to the Russian chemist Dmitri Ivanovich Mendeleev, who began drafting it as an organizational scheme for a textbook he was writing around 1870. He noticed that there were certain recurring patterns in the properties of the known elements when they were placed in order of increasing

---

* This includes many obvious things—periodic tables of meats and craft beers—and some things that wouldn't necessarily occur to most people. Among the odder items in a summer 2017 Google search for unusual periodic tables are a "Periodic Table of Stretching Exercises" and a "Periodic Table of Bank Regulation & Compliance."

atomic mass. The highly reactive alkali metals (lithium, sodium, potassium), for example, are each separated by sixteen to seventeen units in mass, as are the alkaline earths (beryllium, magnesium, calcium), with each alkaline earth one to two units of mass heavier than the corresponding alkali (beryllium has an atomic mass of nine units compared to lithium's seven; magnesium twenty-four to sodium's twenty-three, etc.). When listed in order of mass, the elements start to fall into the familiar rows and columns, with chemically related elements recurring at intervals of eight elements for lighter atoms, and eighteen for heavier ones. Most importantly, Mendeleev used his scheme to predict the properties of then-unknown elements that fell into gaps in his table. The subsequent discovery of the elements scandium, gallium, and germanium, with properties matching Mendeleev's predictions, secured his reputation as the inventor of the periodic table.[*]

As with many scientific breakthroughs in the late nineteenth century, though, the empirical success of Mendeleev's scheme brought with it a nagging problem. The periodic relationship he noticed was undeniably present, but nobody knew *why* this should be the case. And there were small hints that the understanding of the periodic law was not complete, most notably the case of tellurium and iodine: chemically speaking, tellurium belongs in the column before iodine in Mendeleev's scheme—its properties are more like those of sulfur, whereas iodine's are more like bromine—but its atomic mass is greater than that of iodine. Mendeleev argued at the time that tellurium's atomic mass had been incorrectly measured—this had happened with some other elements, such as beryllium—but further experiments only confirmed that tellurium is indeed the heavier of the two. The tellurium-iodine problem was a sign that atomic mass is only a proxy for the real ordering by atomic number, an issue that was not to be resolved for another forty years.

---

[*]   The French geologist Alexandre-Émile-Béguyer de Chancourtois and German chemist Julius Lothar Meyer also came up with periodic classifications of the known elements, but Mendeleev was the only one to use "gaps" in his table to predict new elements, which is why he gets most of the credit for inventing the modern periodic table.

One of the chief properties figured into Mendeleev's scheme was the "valence" of a particular element, which is (somewhat loosely) the number of bonds an atom of a given element can form with other atoms. Starting in the early 1800s (with the work of English chemist John Dalton, refined by Amedeo Avogadro), chemists had noticed that in simple molecules, elements combine in fixed proportions—two units of hydrogen combine with one of oxygen to form water, and three units of hydrogen combine with one of nitrogen to make ammonia. This "law of proportions" was one of the strongest arguments for what became modern atomic theory. As the century went on, this was extended to the idea of a maximum number of bonds per element, a property shared by all elements in a given column of Mendeleev's table. Thus, the alkali metals in the first column all form a single bond, while carbon and its fellows in the fourteenth column of the table (silicon, germanium, tin, and lead) can each bond to four other atoms. Like other chemical properties, the valence repeats after every eight elements for relatively light atoms, and eighteen for heavier atoms.

In the decades following Mendeleev's table, numerous discoveries began to provide clues as to the underlying structure of atoms that results in their periodic behavior. When Mendeleev was putting together his table, the electron had not yet been identified; with the 1897 demonstration that the electron is a particle found within atoms, physicists and chemists began to consider its role in the formation of bonds. The development of Rutherford's solar system atomic model, where the outer portion of the atom is made up of orbiting electrons, suggested a connection between these electrons and the number of bonds. Niels Bohr's model of a limited set of allowed orbits led to the idea of "electron shells," each able to hold a limited number of electrons. In the shell picture, developed by American chemist Gilbert Lewis around 1916, bonds are formed by the exchange or sharing of electrons in order to provide each atom with a completely filled outermost shell.

The arrangement of electrons within atoms also proved key to fixing the problem of ordering atoms in Mendeleev's table. In Bohr's model, the energy of electron orbits is determined by the electromagnetic interaction between the electron and the nucleus, an interaction that gets stronger as the charge of the nucleus increases. This relationship between charge and energy was confirmed by Rutherford's student Henry Moseley in studies of the x-rays emitted by particular elements. While the full pattern of x-ray lines emitted by any element is rather complex, Moseley found that the longest-wavelength x-rays emitted by each element followed a simple pattern, with those wavelengths getting shorter (and the frequency getting higher) as he moved up through the periodic table. Bohr's atomic model offers a simple interpretation of these x-rays as resulting from a transition between the two lowest-energy states of a multi-electron atom, and predicts that the energy of these x-rays should depend on the square of the charge on the nucleus, a prediction that was perfectly matched by Moseley's data.

Moseley made a systematic study of as many substances as he could manage, and he showed that the measured energies fit Bohr's system very nicely for all the elements whose place in the periodic table was well understood. This established x-ray spectroscopy as a way of directly determining the charge of the nucleus—that is, the number of protons present—and established nuclear charge, rather than atomic mass, as the correct method for ordering atoms in the periodic table. This cleared up the mysterious "reversals" of elements, like iodine and tellurium, where chemical properties suggested they go in a different order than atomic weight: tellurium with fifty-two protons must come before iodine with fifty-three. Because protons make up much of the mass of an atom, nuclear charge often closely corresponds to atomic mass, but not entirely: the extra mass of tellurium is due to an additional neutron, a particle that wasn't discovered until 1932 by yet another Rutherford colleague, James Chadwick.

Very much in the spirit of Mendeleev, Moseley also used his results to identify "gaps" in the table to be filled by new elements, all of which were later found, at atomic numbers 43 (the radioactive element technetium), 61 (the radioactive element promethium), 72 (hafnium), and 75 (rhenium). Alas, Moseley did not live to see the confirmation of his work, as he was killed during the battle of Gallipoli in August of 1915.[*]

Moseley had established a method of measuring the number of positively charged protons in the nucleus, which for a neutral atom must be balanced by an equal number of electrons. And by the early 1920s, it was well established that the chemical nature of elements is determined by electron "shells" containing multiple electrons of the same energy, with a maximum capacity for each shell. The shells map to rows of the periodic table—the first and innermost shell can hold up to two electrons, corresponding to hydrogen and helium, which each have only the one shell with one and two electrons, respectively. The next two shells can each hold another eight, accounting for the second (lithium, beryllium, boron, carbon, nitrogen, oxygen, fluorine, and neon) and third (sodium, magnesium, aluminum, silicon, phosphorus, sulfur, chlorine, and argon) rows. The next two shells after that hold eighteen electrons each, then two more each hold thirty-two.

The idea of electron shells connects naturally to Bohr's notion of discrete atomic states, but why the stationary states of the Bohr model should have any limit on the number of electrons they could hold—let alone the observed sequence of capacities two, eight, eight, eighteen, eighteen, thirty-two, thirty-two—remained a mystery. Some physicists attempted to connect it to geometry, noting that eight is the number of corners on a cube, but this didn't go very far. Understanding the origin of chemical structure would require digging deeper than the original Bohr model.

---

[*]   Despite the efforts of many friends and colleagues to keep him in the lab, Moseley felt it was his duty to fight, and enlisted when World War I broke out in 1914. Following his death, Rutherford and others used that tragedy to argue that promising scientists should be kept off the front lines, and serve instead in a technical and research capacity. This arguably laid the foundation for the massive scientific efforts of WWII, leading to the development of radar and atomic weapons.

FROM "OLD QUANTUM THEORY" TO MODERN QUANTUM MECHANICS

There's an old joke in science circles about a dairy farmer, who—applying the sort of logic that only makes sense in jokes—consults a theoretical physicist about how to get more milk from his cows. After a few days, the physicist announces that he has the solution, and the excited farmer asks to hear it, only to have the physicist begin with "First, we assume a spherical cow . . ."

Like most old jokes, this is funny because it captures something true—in this case, about the way physicists operate. The first step in a physics approach to any problem is to reduce it to the simplest case imaginable, even when this means treating complicated objects like cows as smooth spheres. At its best, this approach allows physicists to develop simple universal principles that illuminate the deep workings of nature. Of course, such exceedingly simple models often miss some details—such as the fact that cows are manifestly not spheres—and require later refinements to capture the complexity of the real world. The art of being a physicist consists of starting with spherical cows, and then adding as few additional complications as possible to produce the simplest model that satisfactorily describes the actual universe.

Bohr's quantum model of the hydrogen atom is a "spherical cow" in the finest tradition. It solves an outstanding problem by proposing a strikingly simple fundamental principle, but it considers only the simplest possible case: that of electron orbits that are perfectly circular. This circular-orbit model was sufficient to let Bohr explain the pattern of spectral lines from hydrogen as resulting from transitions between a set of states whose energy was described by the value of a single "quantum number": n. The original Bohr model couldn't capture all the complexity of real atoms, though, such as the "fine structure" of hydrogen (where some spectral lines turn out to be pairs of very closely spaced lines), or the way that single spectral lines split into multiple lines when atoms are placed in a magnetic field.

Bohr's model clearly had the right general idea, but it needed to expand to encompass additional complexity by adding additional states, and the assumption of circular orbits was an obvious place to attack. Within a few years of Bohr's initial model in 1913, Arnold Sommerfeld found a new way to express Bohr's quantum condition that allowed for the existence of elliptical orbits. This led to a richer set of permitted electron states, each described by three integers: Bohr's n, and two new ones, which we'll call l and m.* These new "quantum numbers" have tight restrictions on their possible values: l must always be less than n, and m ranges between a maximum of $+l$ and a minimum of $-l$.

In physical terms, l describes the eccentricity of the elliptical orbit—larger l values are more circular—and m describes how that orbit is tilted. The maximum positive value of m for a given n and l corresponds to an electron in a counterclockwise circular orbit when seen from above, while a negative m corresponds to a clockwise orbit. An orbit with $m = 0$ is a circle standing on end, orbiting up and down.

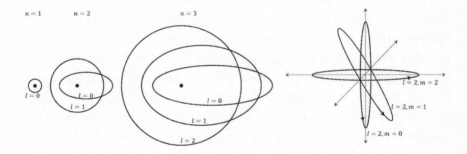

Electron orbits from the Bohr-Sommerfeld model: on the left, some orbits showing the effect of changing n and l. On the right, three orbits for the n = 3, l = 2 state, showing the tilting of the orbit with changing m values.

This Bohr-Sommerfeld atom, which was the dominant model of the "old quantum theory," turns the single allowed energy states of the Bohr model into groups of states with very similar energies—this turns

---

* Using l and m for these is a little ahistorical—the actual numbers used in Bohr-Sommerfeld theory had different names and were interrelated. The analogous quantities in modern quantum theory are l and m, though, so for simplicity I'll use the more modern symbols, and give the correct interpretation later.

out to be exactly what's needed to explain the phenomena that can't be captured by the original Bohr model. Maybe the greatest triumph of the Bohr-Sommerfeld model was explaining the fine structure of hydrogen. Sommerfeld's first atomic model found energies that depended only on Bohr's original quantum number n, but when he incorporated Einstein's special relativity into his model, he found a small energy shift that depended on the quantum number l. Electrons move at very high speeds, just under 1 percent of the speed of light, fast enough that their energy must be calculated relativistically. For circular orbits, the electron's speed does not change, but in elliptical orbits, it speeds up and slows down, so its (relativistic) kinetic energy changes; the resulting shift between the two different l values for the $n = 2$ state matches the fine structure splitting.

Even after adding relativity, though, the quantum number m had no effect at all on the energy of electrons in an isolated atom: its contribution is in describing the *change* in energy when an atom is placed in a magnetic field. The orbiting electron can be thought of as a tiny loop of current, which will behave like an electromagnet with the direction of the north pole determined by the direction of the orbit. When placed in a magnetic field, an electron in the maximum m state orbiting counterclockwise will increase its energy slightly, an electron in the minimum m state orbiting clockwise will decrease its energy slightly, and the $m = 0$ state does not change. This separation between m values explained the "Zeeman effect," where an applied magnetic field splits a single spectral line into three closely spaced lines whose separation increases with the strength of the field. To a point, anyway—the Bohr-Sommerfeld scheme could handle the ordinary Zeeman effect, but not the "anomalous Zeeman effect," where the lines split in two. This remained a vexing problem—in one famous story, a colleague meeting Wolfgang Pauli on the street remarked that he was looking glum, to which Pauli replied, "How can one look happy when he is thinking about the anomalous Zeeman effect?"

The end result of the addition of l and m values, whether in the Bohr-Sommerfeld model or modern quantum mechanics, is that atoms

have groups of "degenerate" electron states* with exactly the same values of n and l, and thus exactly the same energy. The ground state of
hydrogen is a single level with $n = 1$, $l = 0$, and $m = 0$, followed by four
states with $n = 2$: one state with $n = 2$, $l = 0$, $m = 0$, and then a group of
three states with $n = 2$, $l = 1$, and m values of -1, 0, $+1$, having exactly
the same energy. Then there were nine states with $n = 3$: one by itself,
a group of three, and a group of five, and so on.

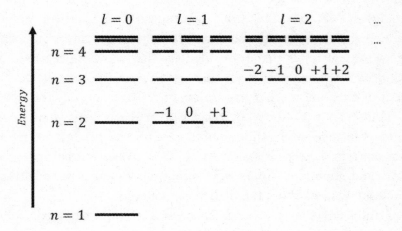

*The degenerate energy levels of the Bohr-Sommerfeld model, showing
the collections of* n, l, *and* m *states with nearly identical energy.*

These collections of degenerate levels suggest some connection
with the electron shells needed to explain the valences of chemical elements, but the numbers of degenerate states (1, 1, 3, 1, 3, 5, and so
on) don't fit the pattern of the periodic table (2, 8, 8, 18 . . .). It's also
not clear how and why the electrons should be distributed among these
states, as opposed to, say, clustering all together in the lowest-energy
orbit available.

These puzzles were solved in 1924 with a bold stroke by Wolfgang
Pauli, who happened to be a former student of Sommerfeld's. After

---

*   The term "degenerate" comes from mathematics and does not imply a moral
     judgment about cohabitating electrons.

struggling to understand the various states of the Bohr-Sommerfeld atom, Pauli realized that a simple trick could reduce the sequence of electron capacities 2, 8, 8, 18, etc. down to a single number: 1.

CHEMISTRY BY EXCLUSION

Pauli was recognized from a young age as something of a physics prodigy. He earned a PhD from the Ludwig-Maximilian University in Munich at just twenty-one, and shortly thereafter wrote a monograph reviewing Einstein's theory of relativity that was long regarded as a definitive work on the subject. He played a crucial role in the development of quantum mechanics through the 1920s, not only through his own direct contributions, but as a key node in an extensive network of correspondence among the community of physicists working on quantum matters. Letters between Pauli, Bohr, Werner Heisenberg, Max Born, and numerous others were important channels for exchanging and refining ideas about the theory as it emerged.

Pauli's PhD thesis was an unsuccessful attempt to describe molecular hydrogen ions—that is, two hydrogen atoms bound together in a molecule, but missing one of the electrons—in the context of the "old quantum theory" of the Bohr-Sommerfeld model. This failure was one of the major factors driving physicists toward the new quantum mechanics, as it convinced them that something must be wrong with the paradigm of well-defined electron orbits. Pauli's deep engagement with the problems of atomic and molecular structure, though, provided the inspiration for his greatest direct contribution to physics.

Considering the many combinations of n, l, and m states available to electrons, Pauli saw that the various electron shell capacities could all be explained at once if you added a *fourth* quantum number, one that could only take on two values. This two-valued quantum number, which we'll call s, would double the number of unique states, after which the electron shell capacities could easily be explained by introducing a new principle (now called the "Pauli exclusion principle"):

each quantum state, defined by a particular combination of n, l, m, and s, can hold one and only one electron.

Thus, hydrogen in its lowest energy state has its lone electron in the lowest Bohr-Sommerfeld level, $n = 1$, $l = 0$, $m = 0$. Helium adds a second electron with the same n, l, and m, and thus the same energy, but the other possible value of the fourth quantum number. Lithium's first two electrons fill those same two states, leaving the third with no choice but to move up in energy to the $n = 2$, $l = 0$, $m = 0$ state. Beryllium takes the other s value for that state, and thus boron's last electron needs to move into the $n = 2$, $l = 1$ trio of states (whose energy is only a tiny amount higher than the $n = 2$, $l = 0$, $m = 0$ state), and so on up through the periodic table.

Pauli exclusion, along with the degenerate states of the Bohr-Sommerfeld model, helps explain chemical bonds, and how atoms form molecules, in terms of electron shells. The alkali metals in the far left column of the periodic table have only one electron in their outermost shell, which they will readily give up to another atom, making them highly reactive. The reactivity of the alkalis increases as you move down the column—if you drop lithium in water, it will fizz mildly, while if you drop cesium in water it will violently explode—because the outer electrons in a heavier atom are farther from the nucleus and less tightly bound. At the other end of the table, the halogens have an outermost shell that contains seven of a possible eight electrons, meaning they will eagerly grab an electron from any other atom to fill that hole. In the halogens, the lighter elements are more reactive—fluorine tops the list of substances that scare chemists, while iodine is mild enough to be used as an antiseptic—because the shell they're trying to fill is more tightly bound.

Carbon sits near the middle of the table, with the $n = 2$, $l = 0$ shell filled and two of a possible six electrons in the $n = 2$, $l = 1$ shell. This allows it to form up to four bonds per atom, making it an extremely versatile element, allowing the enormous variety of organic molecules responsible for the odors and flavors of our food. The bonds formed by carbon are not all that strong, though—it needs to gain or lose four electrons to have a full outer shell, so no one bond is all that important—making it relatively easy to break organic molecules apart and rearrange the pieces, which is why carbon chemistry is the basis for all known life. Silicon, just below carbon in the periodic table, can also form four bonds per atom, and it's regularly suggested as an alternative basis for life, but the bonds silicon forms are a bit stronger than those in carbon—this is probably what keeps silicon-based organisms confined to the realm of science fiction.

SPIN

Pauli's bold proposal was very much in the tradition of Planck, Einstein, and Bohr, introducing an ad hoc element to explain an otherwise mysterious phenomenon.[*] Like those earlier ideas, Pauli's exclusion principle offers a certain conceptual elegance: allowing each state to be occupied by only a single electron is a bright-line rule that explains a lot of physics with a minimum of overhead. Of course, like those other ideas, it was also lacking in any apparent physical justification. That is, there was no obvious property of the electron that might correspond to the two possible values of the quantum number s. As it turned out, though, the physical property in question had already been observed two years earlier, in 1922, though the experimenters had no idea what they had done at the time.

Otto Stern and Walther Gerlach were young research assistants in Frankfurt who were inspired to test the quantum atomic theory using the magnetic properties of silver atoms. In the Bohr-Sommerfeld model, the lowest energy state of a silver atom involves an electron moving in a circular orbit with a single unit of angular momentum. As mentioned above, this should behave like a tiny loop of current, making

---

[*]   This was a trick he resorted to a second time, in 1930, when he did "a terrible thing" by proposing the existence of an "undetectable" particle, the neutrino, as a solution to a problem relating to beta decay of nuclei, as described back in Chapter 1.

the atom a little electromagnet with a north pole pointing in a direction determined by the value of m.

Stern doubted Bohr's model and realized that this "space quantization" ought to provide a way to test it. The two directions of orbital motion should lead to two possible magnetic states for the atom, and if a beam of such atoms was sent through a magnetic field that varied in space, the two orientations should separate. Atoms with one orientation lower their energy when a magnetic field is applied, so they're pulled toward the region where the field is strongest; atoms with the other orientation increase in energy in the magnetic field, and so are pushed away from it. Stern designed and Gerlach carried out an experiment to test this using a beam of silver atoms emerging from a small hole in a hot oven inside a vacuum chamber. They passed this beam between the poles of a tapered magnet and onto a glass plate, and then looked for the image produced by the deposited silver atoms. After more than a year of experimental struggles, they managed to get the result they sought: with the magnet in place, their single beam of silver atoms split into two. According to Stern, when they looked at the first plate, they didn't see anything until after he breathed on it, at which point the deposited silver darkened and became visible. He attributed this to a reaction with the sulfur in the cheap cigars he smoked because he couldn't afford better on his salary as a young professor. An excited Gerlach sent a postcard to Bohr with a photo of their data, congratulating him on the confirmation of his theory.

The reaction to the Stern-Gerlach experiment from the physics community, though, was less elation than confusion. While the idea of different orbital directions made sense, it was not clear why the atoms should happen to be equally divided between "up" and "down" states, as opposed to distributed randomly in all different orientations—in which case the beam should smear out, rather than split. The more refined treatment of the Bohr-Sommerfeld quantum model also predicts that the beam should split into *three* inside the magnet, corresponding to the three different values of m; Stern and Gerlach had forgotten to consider the $m = 0$ state when designing their experiment.

Explaining how the Stern-Gerlach experiment could give two and only two beams was a major puzzle.

A quantum property with only two possible values, though, was exactly what Pauli's exclusion principle needed. In 1925 a pair of Dutch physicists, George Uhlenbeck and Samuel Goudsmit, suggested the modern interpretation: that the electron has an intrinsic angular momentum as if it were a tiny spinning ball, and that this "spin" can take on only two values, traditionally described as "up" and "down." This spin angular momentum also gives the electron some magnetic character (which we'll discuss more in Chapter 9), which manifests in the "anomalous Zeeman effect" that had bedeviled Pauli: when an atom in an $l = 0$, $m = 0$ state is placed in a magnetic field, one of the spin states shifts up in energy, the other down, causing spectral lines involving that state to split in two. The same energy shift is what caused Stern and Gerlach's beam of silver atoms (whose outermost electron happens to be in an $l = 0$, $m = 0$ state) to split into two.

This spin angular momentum has some unusual properties, starting with the fact that the magnitude of the electron spin is one-half the fundamental unit of angular momentum used in the Bohr model—the quantum number s takes on values of $s = \frac{1}{2}$ or $s = -\frac{1}{2}$, unlike all the other quantum numbers, which are integers. Notably absent from the possible values is 0, meaning that the electron is never *not* spinning, or spinning about an axis perpendicular to the measurement axis. And the physical nature of the spin angular momentum defies classical explanation—Pauli himself had quashed a similar proposal from Ralph Kronig, a visiting PhD student, a few months before Uhlenbeck and Goudsmit published their theory, on the reasonable grounds that, given its tiny mass and size, if the electron were literally a spinning ball of charge, a point on its surface would need to be moving many times faster than the speed of light to generate the necessary angular momentum. Pauli was infamously acerbic, and not shy about expressing objections to theories that he didn't feel held up. Kronig's spin proposal was dismissed with "It is very clever but of course it has nothing to do with reality." This was relatively mild by Pauli's standards—his most famous criticism being "This is not even wrong."

As with the dual particle-and-wave nature of light, these odd quirks of spin were eventually adopted as fundamental properties of the emerging quantum physics that simply needed to be accepted. The electron is not *literally* a spinning ball of charge, but it carries intrinsic angular momentum as if it were, and that's just how electrons work. The never-not-spinning nature of the electron is also a strange state of affairs, but it's exactly what is needed to explain the Stern-Gerlach experiment.

After being converted from his initial skepticism regarding spin, Pauli played an instrumental role in working out its mathematical description in terms of matrices. In 1930, English theoretical physicist Paul Dirac finally showed that electron spin is an inevitable consequence of combining quantum mechanics with Einstein's special relativity. Even before Dirac's full theory, though, electron spin and the Pauli exclusion principle simply explained too many phenomena (some of which we'll discuss in upcoming chapters) for them *not* to be accepted, however weird and even inexplicable they might have seemed.

FROM ORBITS TO PILOT WAVES TO PROBABILITY

Along with its oversimplicity—its status as a spherical cow—the other outstanding problem with the Bohr model was the seeming arbitrariness of the quantum condition for allowed orbits. That is, why *should* only integer multiples of the angular momentum be allowed in the first place? Sommerfeld's extension of the theory provided a richer variety of orbits, but they still lacked a convincing basis in some physical property of the electrons inside an atom.

The first step toward an answer to this problem came from a French graduate student from an aristocratic family, Louis-Victor-Pierre-Raymond de Broglie (generally just shortened to "Louis de Broglie"), who picked up on a connection to the *other* branch of emerging quantum theory, regarding the nature of light. In his PhD thesis, de Broglie suggested a parallel between light and matter: if light waves have particle nature, then maybe a particle like an electron should have an associated wave, with the same inverse relationship between the wavelength and momentum seen for light; thus, doubling the electron's momentum should cut its wavelength in half. In this wave picture, the Bohr-Sommerfeld quantum model takes on an obvious physical meaning: If you trace out the electron wave around one of the "stationary state" orbits with the principal quantum number n, when you get back to the starting point, the wave has oscillated n times. The allowed orbits are those for which the electron wave wrapped around the orbit forms a standing-wave pattern just like the standing-wave modes of light used to set up the black-body problem back in Chapter 2.

The idea of electrons as waves was an extremely bold suggestion —one popular story says that de Broglie's PhD committee had no idea what to make of his thesis, until Einstein was invited to weigh in and declared it "a first feeble ray of light on this worst of our physics enigmas." Luckily, it is also an eminently testable idea, and within a few years direct experimental evidence emerged. In the US, Clinton Davisson and Lester Germer saw wave diffraction in a beam of electrons bouncing off a crystal of nickel, a discovery made entirely by accident. While they were running their experiment, a break in their vacuum system let in air that oxidized their nickel sample. To clean the surface, they heated it to a high temperature, which caused it to partially melt, and as it cooled it formed much larger crystals, leading to more dramatic (and thus more easily observed) diffraction peaks. Visiting the UK some years later, Davisson was surprised to hear Max Born citing his odd experiment as evidence of the wave nature of electrons. Subsequent experiments confirmed this explanation, though, and Davisson shared the 1937 Nobel Prize for discovering the wave nature of the electron with George Thomson, at the University of Aberdeen, who'd observed diffraction of electrons sent through thin films of grease. Thomson's father was J. J. Thomson, who won the 1906 Nobel Prize for showing that the electron is a particle (see Chapter 3)—dinner-table conversation in the Thomson family must have been interesting.

These experiments showed that, as strange as it seems, electrons *really do* behave like waves, and a radical break from classical physics was needed to accommodate this.

The original scenario envisioned by de Broglie involved the electron as a particle guided by an associated "pilot wave." The mathematics of this never quite worked in the 1920s, and de Broglie eventually abandoned that approach (though the pilot wave idea was revived in the 1950s by David Bohm, an American physicist, and remains an active topic of research).[*]

Inspired in part by de Broglie's proposals, in 1926 the Austrian physicist Erwin Schrödinger developed a wave equation that correctly described the behavior of electrons. Schrödinger's equation was undeniably a great success and won him a share of the 1933 Nobel Prize in Physics, but it has some mathematical peculiarities. In particular, the equation explicitly includes the imaginary number i: the square root of -1.

If you recall learning about square roots in middle school, or have ever tried to figure the square root of a negative number on a calculator, this may seem like an impossibility. In fact, though, it's possible to expand the basic ideas of mathematics to include i as a distinct number, and combining "real numbers" like 1, 2, π, and $\sqrt{2}$ with multiples of i gives us some powerful techniques for analyzing a great deal of physics. These complex numbers are particularly useful in the study of waves and optics, so in one sense it is only natural that they show up in Schrödinger's equation for electron waves.

In the study of classical waves—like light and sound—imaginary numbers are used mostly for calculational convenience: the actual measurable waves are all described by real numbers. The waves in Schrödinger's equation, on the other hand, can only be described by complex numbers with an imaginary component. This means these

---

[*]   The de Broglie-Bohm approach remains something of a niche topic, though, as the decades of neglect between de Broglie's original work and Bohm's reinvention saw huge advances in the more orthodox interpretations, leaving pilot-wave fans with a lot of catching up to do.

wavefunctions can't be describing a real disturbance in some medium, like the ripples on the surface of water. But then the question is, what *are* they describing?

The modern approach to understanding the wavefunctions described by Schrödinger's equation was introduced by Max Born in a footnote to a 1926 paper, and it interprets the wavefunction as related to the probability of finding the electron at a given point. The wavefunction itself is not the probability, because it's a complex number and there are no imaginary probabilities. Instead, the probability is given by the "squared norm" of the wavefunction, a process similar to squaring the wavefunction, but in a way that eliminates the possibility of getting a negative result from the imaginary components.

Solving the Schrödinger equation for an electron in a hydrogen-like atom still gives a discrete set of states labeled by three integers n, l, and m, but the interpretation of these states as probabilities destroys the Bohr-Sommerfeld image of an electron moving along a regular classical orbit. Instead, the wavefunction describes an "orbital," a sort of fuzzy ball of probability surrounding the atom. Any individual measurement of the position of an electron with a particular n, l, and m will find it somewhere in the vicinity of the nucleus, and when repeated an enormous number of times with identically prepared atoms, the measured positions will trace out a probability distribution that's determined from a wavefunction satisfying Schrödinger's equation. The electron in an orbital can't be said to have a well-defined position or momentum; all that can be defined is the probability of observing it at a particular position, or moving at a particular speed. (This has profound consequences for our understanding of physics that we'll discuss more in the next chapter.)

There are, however, some properties for an electron that are well-defined, the most important of which is its total energy. This is still primarily determined by the "principal quantum number" n, which represents the overall energy of the orbital, with energies quite close to those predicted by the Bohr-Sommerfeld atom. The integer n is no longer associated with the angular momentum of the electron in its orbit, though—that role is filled by the quantum number l, which determines the total angular momentum for a given orbital (and, as you can see in the figure below, this relates to the number of nodes). The quantum number l can take on a range of values, but the value of l must always be less than n. Finally, the quantum number m gives the value of the angular momentum along a particular axis; as in the Bohr-Sommerfeld atom, the energy depends only weakly on l, and not at all on m unless there is an applied electric or magnetic field. The Schrödinger equation, then, leads to the same collection of degenerate energy levels found in the Bohr-Sommerfeld model.

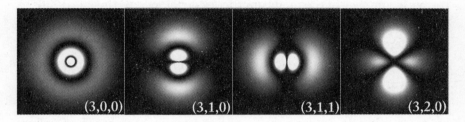

(3,0,0)   (3,1,0)   (3,1,1)   (3,2,0)

Probability distributions for some selected (n, l, m) states for an electron in hydrogen, showing the effects of changing the angular momentum quantum numbers: increasing the value of l gives a pattern with more nodes, while increasing m rotates the pattern. Images are two-dimensional slices through the probability distribution; brighter white indicates higher probability.

Unlike the original Bohr model with electrons orbiting along a plane, though, or the tilted orbits of the different m values in the Bohr-Sommerfeld model, the orbitals calculated from the Schrödinger equation are intrinsically three-dimensional. While the different values of l and m don't have different energies, they do profoundly alter the distribution of the electron in space, as seen in the figure. This provides the last piece needed to understand the structure of molecules. These electron distributions determine the shape and vibrations of molecules, which in turn determine just about everything else about them. The four bonds formed by carbon, for instance, naturally arrange themselves at the corners of a tetrahedron, leading to the three-dimensional structure of many organic molecules. The two bonds formed in oxygen lie in a plane, separated by an angle of about 104 degrees, producing the characteristic chevron shape of water molecules, which is essential to many of its chemical properties, including the unusual fact that water expands when it freezes.

MODERN CHEMISTRY

A physicist from 1917 transported forward a hundred years would have a difficult time recognizing large swaths of the field, thanks to the transformative impact of the quantum theory developed in the late 1920s. A chemist of the same era, though, would find it much easier to get up to speed here in the twenty-first century, as many of the concepts underpinning the modern approach would be familiar. You can still go a long way toward understanding chemical bonding and structure simply by thinking in terms of atomic valence and the sharing of electrons among atoms to fill their electron shells, without really worrying about where those shells come from.

The details, though, are thoroughly quantum. The vague conceptual "shells" of early twentieth-century chemistry have been replaced with quantum-mechanical orbitals whose actual size and shape can be calculated to high precision using the Schrödinger and Dirac equations (the latter is most important for heavier atoms, where the electrons are moving fast enough for the effects of relativity to become significant).

Molecular bonds form as atoms come close enough for their orbitals to begin to overlap and merge together, with the electron wavefunction spread over a wider region encompassing both atoms. The individual electrons involved in a bond, then, are quite literally shared between two atoms (as we'll see in the next chapter).

The wavefunctions of electrons determine not only the precise three-dimensional arrangement of the electron orbitals but also how many states are available within a particular collection of atoms. Pauli exclusion sorts the electrons between those states, minimizing the total energy while ensuring that each electron has a unique set of quantum numbers. This, in turn, determines how quickly and how strongly bonds form between atoms, and how the resulting molecules are shaped. Those structures and bond strengths then determine how any given molecule will react with others, including the chemical receptors in our nose that we use to process a collection of airborne organic molecules into the sensation of smell.

Many problems in chemistry remain difficult—calculating the structure of even simple molecules can be a difficult computational challenge, and even the best supercomputers struggle with finding the shape of large protein molecules involving hundreds of individual atoms in long chains that can coil up to form elaborate three-dimensional structures. And, as we mentioned back at the beginning of this chapter, the details of how odor molecules are detected and processed have still not been entirely worked out. We know with certainty, though, that when you start your morning by enjoying the pleasant aroma of your favorite hot beverage, that drink ultimately gets its smell from Pauli exclusion and the wave nature of electrons.

CHAPTER 7

SOLID OBJECTS: THE ENERGY OF UNCERTAINTY

*I slide a couple of **slices of bread** into the toaster oven, jiggling the rack when it sticks a bit, and **lean against the counter** while I wait . . .*

Almost 2,500 years ago, the philosopher Zeno of Elea published a large number of paradoxes, attempting to demonstrate the absurdity of "common sense" ideas about reality. One of the most famous purports to show that motion is an illusion, because all motion should take infinite time. In order to walk across a room, for example, I first must walk halfway across the room, which takes some amount of time. Then I need to walk half of the remaining half (that is, to the three-quarters mark), which also takes some amount of time. And then I need to walk half of the remaining distance (that is, from three-quarters to seven-eighths), which takes some time as well. This process can continue forever, leading to the conclusion that moving any distance at all requires an infinite number of steps, each taking finite time. Which suggests that it should take an infinite amount of time to move anywhere, and thus motion is impossible.

Most people react to this in more or less the same way as that attributed to Diogenes the Cynic: by standing up and walking away from the philosopher. Motion is such an obvious fact of our everyday existence that declaring it impossible seems ridiculous. More mathematically inclined thinkers address the paradox by pointing out that each time you halve the distance, you also halve the time needed to cross it. With the invention of calculus, we know that an infinite number of successively smaller terms can sum to a finite total (and more specifically, that the sum involved in the setup of this paradox is $\frac{1}{2} + \frac{1}{4} + \frac{1}{8} + \ldots = 1$). Philosophers, however, have continued to argue about subtle points of Zeno's arguments to the present day.

The solidity of objects is another inescapable fact of our existence, something we experience every time we place one object atop another. Questioning the stability of solid matter, then, might seem like the sort of thing best left to philosophers or those who have had a bit too much of some controlled substance. Yet it turns out to be extremely difficult to *prove*, based on the principles of physics, that solid objects consisting of large numbers of interacting particles are actually stable.

The problem is most easily described in terms of energy: as a general matter, any physical system always tries to reduce its energy, so in order for a large number of particles making up a solid to be stable, there must be some minimum energy arrangement from which no further energy can be extracted. As we saw when we talked about the Bohr model, though, the attraction between positive and negative charges gives them a negative potential energy that dives to negative infinity when the two are right on top of each other. That infinite value suggests that any collection of particles can, in principle, lower its energy by packing all the components more closely together. It's not immediately obvious, then, that the attractive interactions that pull particles together to make atoms, molecules, and solids can't, under the right circumstances, pull all those particles into an infinitesimally small space, causing would-be solids to implode—and releasing an enormous amount of energy in the process. A slice of bread waiting to be toasted, in this view, is a potential atomic bomb.

Preventing this implosion to allow for the existence of solid objects requires some additional factor that increases the energy as particles pack more tightly. This would ensure a minimum energy at some size in much the same way that balancing the electromagnetic force against the force needed to hold an electron in a small orbit let Bohr find an optimum radius for the lowest-energy orbit in his atomic model. In the end, this energy comes from two core quantum ideas we've already discussed, specifically the wave nature of matter and the Pauli exclusion principle. To explain how, we'll have to introduce one of the most famous consequences of quantum physics, something no book on the subject can get away without mentioning: the Heisenberg uncertainty principle.

THE CERTAINTY OF UNCERTAINTY

In the mid-1920s, confronted with the inability of the Bohr-Sommerfeld model to correctly describe some seemingly simple systems—like the ionized hydrogen molecule that Wolfgang Pauli struggled with during his PhD—a number of (mostly) young physicists began to abandon the semiclassical underpinnings of the "old quantum theory." The first breakthrough came when Werner Heisenberg decided that the key to the problem was to dispense with the idea of well-defined electron orbits altogether.

Heisenberg was of the same generation as Pauli (who was a year older), and like Pauli studied under Arnold Sommerfeld in Munich. His thesis research was on the classical physics of turbulence, but like many physicists of the day, he developed an interest in the emerging quantum theory. After he completed his PhD, he moved to Göttingen to work with Max Born, and spent the winter of 1924–25 at Niels Bohr's institute in Copenhagen. While in Denmark with Bohr, he tried to use "old quantum theory" to explain the intensity of spectral lines—that

is, why atoms emit and absorb more readily at some of their characteristic frequencies than others. Einstein's statistical model of light from
Chapter 3 gave some very general rules for these, but getting the details
right turned out to be exceedingly difficult.

In the summer of 1925, Heisenberg returned to Göttingen and
continued to struggle with the problem of spectral lines. At the same
time, he was also struggling with allergies, and to escape an intense
bout of hay fever, fled to the remote island of Heligoland, where he
could breathe pollen-free air and concentrate on his work. While there,
he had an epiphany, realizing that it was a waste of time to try to figure
out details of the classical orbits followed by electrons. No conceivable
experiment could hope to track the motion of the electron in its orbit,
so there was no point in worrying about the finer points of that motion.
Instead, he looked for a way to formulate quantum theory solely in
terms of experimentally observable quantities.

After an extended period of furious mathematical effort, Heisenberg found the answer he was seeking. This took the form of laboriously calculated tables of numbers describing the measurable
properties of quantum jumps; these values were sorted into rows
and columns based on the particular pairs of initial and final electron
states involved. As with Schrödinger's wave equation in the previous
chapter, Heisenberg's theory deals in probabilities—but in terms of
allowed states, rather than particular positions for the electron. The
problem of the intensity of spectral lines that motivated Heisenberg
was a matter of determining the probability of an electron in one
allowed state making a jump to another; the higher the probability,
the brighter the line. Making these calculations involved combining
results from his tables of numbers, according to rules that he slowly
worked out.

On returning to Göttingen, Heisenberg showed his work to Born,
who noticed a similarity between Heisenberg's calculations and the
matrices—indexed tables of numbers with special rules governing their
manipulation in calculations—studied by colleagues in mathematics.
Born and Heisenberg and another of Born's assistants, Pascual Jordan,
reformulated Heisenberg's results in the language of matrices, leading

to the first relatively complete theory of quantum physics: "matrix mechanics."*

The initial reception of matrix mechanics was not all that enthusiastic, as physicists of the day were mostly not trained in the mathematics of matrices; when Erwin Schrödinger found his wave equation the following winter,† many physicists reacted with relief. The two approaches are mathematically equivalent, though, and these days physicists learn a mix of both. The wavefunctions calculated with the Schrödinger equation are regularly described using mathematical terms borrowed from matrix mechanics, while Heisenberg's insights are often described in terms of waves when that offers a more intuitive way of understanding what's going on. Actual calculations are done using whichever approach is easiest for the problem at hand.

Heisenberg is best known outside of the physics community for his uncertainty principle, which is one of the ideas from quantum physics that has escaped into popular culture. The most famous formulation of this says that it is impossible to know both the position and the momentum of a particle to arbitrary precision. Both of these quantities are necessarily uncertain, and the product of their uncertainties must be greater than some minimum value—in other words, when uncertainty in one decreases, the uncertainty in the other must increase by at least the same factor. If you know exactly how fast something is moving, you lose your ability to know where it is, and vice versa.

The uncertainty principle is often described as a measurement phenomenon, with the act of attempting to measure the position disturbing the momentum, and vice versa. While this does get to the right basic

---

*   Heisenberg won the 1932 Nobel Prize for developing matrix mechanics, though he wrote to Born afterward that he felt guilty over this, as the work had been done jointly by the three of them. Born eventually received the 1954 Nobel Prize; the delay is frequently attributed to politics, as Jordan was an early and enthusiastic supporter of the Nazis, and it took a while to find a way to honor Born without also including Jordan.

†   Like Heisenberg's, Schrödinger's breakthrough came while he was away from home, in his case on a ski holiday with one of his many mistresses. Ever since, physicists have tried to use these examples to argue for more vacation time.

relationship, it's slightly misleading in that it leaves the impression that there is a "real" position and momentum associated with a quantum particle, and we just don't know what it is. Quantum uncertainty is more fundamental than that, though. One of the words Heisenberg used when working out his theory is arguably better translated as "indeterminacy" than "uncertainty," which offers a more useful way of thinking about the issue. Quantum indeterminacy follows directly from Heisenberg's original epiphany on Heligoland: the idea of formulating the theory only in terms of measurable quantities implies the existence of other quantities that cannot be determined. The uncertainty principle isn't about deficiencies in measurement—it reflects the fact that it simply does not make sense to speak of a well-defined position or momentum for a quantum particle.

To understand this indeterminacy, though, and how it helps produce the energy we need to keep our breakfasts from imploding, we must first circle back and make use of Schrödinger's wave picture. We need to look more closely at what it means for a particle to behave like a wave, and vice versa.

ZERO-POINT ENERGY

The Heisenberg uncertainty principle is the best known of the weird consequences of quantum physics, but to explain it, we need to look at another phenomenon that's just as deeply counter to our intuition. That idea is "zero-point energy," which tells us that a confined quantum particle is never *not* in motion, and it follows directly from the wave nature of quantum particles. This will turn out to have profound consequences for the stability of matter.

To gain some insight into wave nature and zero-point energy, it's useful to go back to the simplest system in quantum physics: a single particle confined to a box.* The basic idea is the same as the "stuff in a

---

* You might think that the simplest possible system would be a particle by itself in free space, but that actually turns out to be much more complicated, as we'll see in a little bit.

box" model we used to set up the black-body radiation problem, but in that case, we were considering light waves. Now, armed with de Broglie's idea of matter waves, we want to consider a material particle like an electron confined to a box. Our hypothetical "box" is impenetrable, so that while the electron can move about the interior freely, it can never escape.

Though the scenarios may seem quite different from the standpoint of everyday intuition, there's very little mathematical difference between a light wave confined to a reflecting box and an electron with wave nature held in an impenetrable box. In both cases, the end result is a limited set of standing-wave modes, with the waves constrained to be zero at the ends of the box, and an integer number of half wavelengths fitting across the box's length. Just as with the light waves, the longest possible wavelength for an electron inside the box is twice the length of the box.

When applied to light waves, this constraint didn't seem problematic, but it has a very unusual implication when applied to an electron: namely, that the electron can never be truly at rest within the box. As de Broglie showed, the wavelength of an electron is related to its momentum—higher momentum means a shorter wavelength. Momentum is calculated by multiplying mass times velocity,* and since the mass of the electron is fixed, the electron's momentum is a reflection of its velocity. An electron at rest would have zero momentum, which would require an infinitely long wavelength. But a confined electron has a maximum possible wavelength—twice the length of the box—which means it has a minimum momentum that is *not* zero. Thus an electron confined to some region of space must always be moving.

In physics, velocity is a quantity that includes both a magnitude (speed) and a direction, and as a result, momentum is also defined in terms of direction. Since an electron in a box can be moving in any direction, though, this makes it tricky to talk about confined particles

---

* This is true for slow-moving particles, anyway; once speeds start to approach the speed of light, relativity changes the definition slightly, but for our present purposes, "mass times velocity" is adequate.

in terms of momentum. To avoid the direction problem, it's easier to discuss the confined electron in terms of its kinetic energy, which does not depend on where the particle is headed, only on how fast it's moving. The standing-wave modes for an electron trapped in a box are states with well-defined kinetic energy, with the energy increasing proportionally to the square of the number of half wavelengths in the box—that is, the second state has four times the energy of the first, the third state nine times the energy of the first, and so on.

The critical feature here is that the lowest energy is *not* zero. This seems like a strange thing to say, from the perspective of classical physics—if I place a macroscopic everyday object like a marble inside a shoebox, I can perfectly well arrange for that object not to move at all relative to the box, and thus to have zero kinetic energy. A quantum particle, on the other hand, can never be perfectly still, thanks to its wave nature. This minimum energy—zero-point energy—is unfortunately a rich source of material for scams. People with a little knowledge of quantum terminology and no scruples will sometimes pitch "free energy" schemes based on the idea of extracting this zero-point energy from empty space. As usual with promises that sound too good to be true, this is impossible: the zero-point energy is simply an inevitable consequence of the wave nature of matter and can never be extracted.

The electron's minimum energy is set by the size of the box, and it depends on the inverse square of the length—that is, if you double the length of the box, the minimum energy of this larger box will be one-quarter that of the smaller. The more tightly you confine a particle, the shorter its maximum wavelength gets—and the higher its energy. This increase in energy is one of the crucial elements we need to understand the stability of matter.

THE UNCERTAINTY PRINCIPLE

The wave nature of matter, then, ensures that a confined particle has some minimum energy, but it's probably not obvious how this relates to the uncertainty principle. Why does the wave nature of matter make it impossible to know the position and momentum of a single particle at a given time?

The key idea is hiding a few paragraphs back, where we switched to talking about the energy of states. The standing-wave states of a confined electron are states with a definite energy but an indefinite momentum, because, as mentioned, the momentum includes not only the speed of the particle's motion, but also its direction. The simplest version of our "particle in a box" hypothetical is a one-dimensional "box"—somewhat like a string, in that the electron is able to move in only two directions. An electron confined to a one-dimensional box is equally likely to be moving either to the left or to the right, giving it an uncertainty in momentum. For a one-dimensional system, we encode the direction of motion into the sign of the particle, so a leftward-moving particle has negative momentum and a rightward-moving particle has positive momentum. The spread in momentum, then, is twice the momentum associated with the fundamental wavelength of an electron. We can express this as an average momentum with some uncertainty: for instance, if the momentum can be either 5 or −5 units, the range of momentum would be 10 units, and we would say that the average momentum is 0, plus or minus 5 units.

More tightly confined particles must have shorter wavelengths, and thus greater momentum and energy, so we can reduce the momentum, and in turn the momentum uncertainty, by increasing the size of the box. When we do that, though, we necessarily increase the uncertainty of the position of the particle, which is something like half the size of the box—on average, the particle is in the middle, and could be up to half the length away in either direction.* The *product* of these two uncertainties, though, is a constant: if we double the length of the box, we double the position uncertainty but cut the momentum uncertainty in half, so position uncertainty multiplied by momentum uncertainty remains unchanged.

---

*   The exact value for a one-dimensional box is a bit smaller than this because the probability distribution is peaked in the middle, but this gets at the basic idea.

So both the position and the momentum of a particle in a box must be uncertain in the way that the Heisenberg uncertainty principle demands. It may not be as obvious, though, that this applies to the case of a particle outside a box, one that's free to move around as it pleases. To understand that, we need to think through what it means for a quantum object to have both particle and wave nature, and what we're asking for when we try to define both its position and momentum.

In order to talk about a quantum particle—that is, a particle with wave nature—having a well-defined momentum, we need to be able to specify its wavelength, which necessarily means it must extend over enough space for us to see it oscillate. But this is not compatible with having a perfectly well-defined position. The best compromise we can make is to have something like a "wave packet," a function where you have wavelike behavior in only a small region of space, as shown in the illustration.

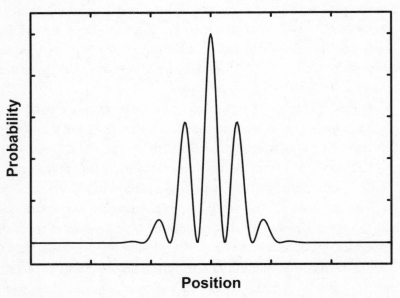

A wave packet, with an obvious oscillation only in a small region of space.

This function clearly has both particle and wave characteristics, but how would we make such a thing out of ordinary waves? We can take a cue from the case of the particle in a box, where the lowest energy state is the sum of two different waves, one corresponding to the particle moving to the left, and the other to the right. Rather than having the particle moving at the same speed in two directions, though, let's look at what happens when we add together waves corresponding to two different possible speeds. In that case, we end up with a wavefunction that looks something like the following illustration.

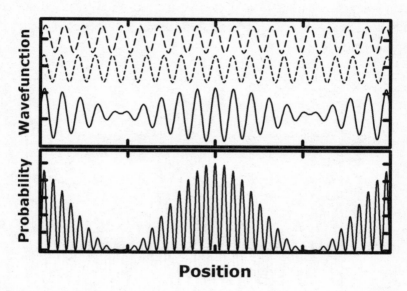

Adding two waves of slightly different frequency gives a wavefunction with beat notes where the two waves cancel out. We square this to get the probability distribution at the bottom.

When two waves with different wavelengths are added together, there are places where the two line up in phase and combine to produce large waves, but as the waves move along, they get out of phase with one another. Some distance away, there will be a point where they cancel each other out almost perfectly, leading to no waves at all. This is referred to as "beating," because it's a familiar phenomenon in music, leading to a discordant ringing noise when two slightly out-of-tune instruments try to play the exact same note.

With only two waves added together, we end up with only narrow regions of no waving, but if we add more waves, the regions where the waves cancel get broader, and the regions where there *are* waves become narrower and more well-defined. The more wavelengths you include, the more the resulting wavefunction resembles a wave packet that describes a particle. Each additional wavelength added, though, corresponds to a possible momentum. As you add wavelengths, you introduce a probability of finding the particle at each particular momentum; you get a smaller wave packet with a more well-defined position, but this process necessarily increases the uncertainty of the particle's momentum.

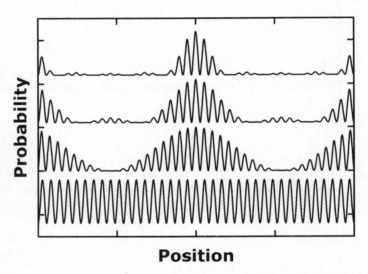

Adding (bottom to top) one, two, three, and five wavelengths to make successively narrower wave packets.

This is why quantum uncertainty is perhaps better described as "indeterminacy": the tension between particle and wave properties means that it's simply impossible to define both the position and the momentum of a particle at the same time. Making a narrow wave packet to better define the position necessarily means adding wavelengths and increasing the uncertainty in momentum. On the other hand, reducing the number of possible wavelengths to better define the momentum will necessarily lead to a wider wave packet, with a greater uncertainty in position. Quantum uncertainty is not a practical limit on our ability to measure things, but a fundamental limit on what sorts of properties a quantum particle can have.

THE STABILITY OF ATOMS

So, how do zero-point energy and the uncertainty principle help ensure the stability of matter? To understand this, we need to discard the simple but rather artificial particle-in-a-box model in favor of the more realistic situation of an electron bound to the nucleus of an atom.

An electron bound to an atom is clearly a more complicated situation than an electron confined to a box, but similar considerations come into play. A bound electron, more or less by definition, is restricted to a small region of space around the nucleus, and just as in the case of the electron in the box, the size of that region determines a minimum kinetic energy that the electron must have.

The case of an atom, however, is complicated by the attractive interaction between the negatively charged electron and the positively charged nucleus. The convention in physics is to describe such an interaction in terms of a bound electron's negative potential energy, which adds with the positive kinetic energy to establish the total energy of the particle. As mentioned earlier, this gives us a simple way to determine whether an electron is bound or not—bound electrons have negative total energy. (This is why the electron energy in a Bohr orbit that we described in Chapter 4 is a negative number.) The law of conservation of energy tells us that this total energy is a constant, with kinetic energy increasing as potential decreases, and vice versa, to keep the sum the same.

While the potential energy of a bound electron is always negative, this energy varies with position. At large distances between the electron and nucleus, it's nearly zero, and as they come closer together, it becomes more and more negative. Mathematically, the magnitude of this negative potential energy increases without limit—placing the electron exactly on top of the nucleus should result in a potential energy of negative infinity. This is what raises the uncomfortable prospect that the atom might be unstable against implosion—that is, that the electron could always lower its total energy by moving closer to the nucleus.

Happily, it's not very difficult to show mathematically that the increase in kinetic energy that comes from confining the electron more tightly is enough to counter the increase in negative potential energy. In fact, this kinetic energy posed a significant historical problem for nuclear physics—atomic masses are always greater than the number of protons inferred from the charge of the nucleus, so prior to the discovery of the neutron, physicists assumed that the nucleus must contain some number of additional protons with tightly bound "nuclear electrons" to cancel their positive charge. The kinetic energy of a confined electron becomes so enormous, though, that it's impossible to hold one inside the space of an atomic nucleus with the interactions known to physics. The "nuclear electron" model never worked all that well, and Ernest Rutherford, among others, believed for many years that the nucleus must also contain a heavy neutral particle. When Rutherford's colleague James Chadwick demonstrated the existence of the neutron in 1932, following up on hints in a paper by Frédéric and Irène Joliot-Curie, many physicists were grateful to be done with "nuclear electrons."

Thanks to the increasing kinetic energy that comes from restricting the electron to a smaller space, the total energy of an electron orbiting a nucleus has a lower limit. The energy of the electron is negative, indicating that it is bound, but it can never be made *infinitely* negative, so its wavefunction must always extend over some range about the nucleus. An atom consisting of an electron bound to a positively charged nucleus is stable, and will not implode.

PAULI EXCLUSION AND SOLID MATTER

The wave nature of matter is enough to guarantee the stability of atoms, so it might seem the philosophical question regarding the existence of macroscopic objects is settled. The fact that a single nucleus orbited by a single electron is stable, however, does not necessarily mean that a large collection of nuclei and electrons will be. The single-atom calculation is simple enough to be a homework problem for undergraduate physics students, but once you add even a third charged particle, it becomes impossible to perform an exact calculation of the energy with pencil and paper; only approximate solutions and numerical simulations are possible.

This is not a problem that's unique to quantum physics. The classical "three-body problem" is similarly intractable, and it was troubling people long before Planck's introduction of the concept of energy quanta. The problem of multiple interacting objects first became a serious concern when Isaac Newton introduced his law of universal gravitation in the late 1600s, and used it to explain the orbits of the planets in the solar system. The basic properties of these orbits can be determined by considering the interaction between a given planet and the sun—but, of course, there are also the gravitational forces between the planets to consider. These are much smaller, but not insignificant: in 1846, the French astronomer Urbain Le Verrier used a minute deviation between the predicted and observed orbits of the planet Uranus to infer the presence of another planet orbiting even farther from the sun. Le Verrier predicted the location of this new planet using approximate calculations with Newtonian gravity, and the German astronomer Johann Galle found the planet Neptune in almost exactly that spot on his first night of observing after receiving Le Verrier's prediction.

Despite the success of approximate orbital calculations like Le Verrier's, the lack of a definite solution to the three-body (or more) problem remained a headache, with troubling implications for human existence. While the forces between individual planets are quite small compared to the gravitational attraction of the sun, if they align in just the wrong way, they could conceivably destabilize the orbits of the planets, flinging Earth into the sun, or out into the depths of interstellar space. In the absence of a definite solution to the many-body problem, there's no guarantee that the solar system will continue to exist in its current configuration.

In 1887, in an attempt to settle the issue, the King of Sweden declared an international competition, with a prize for any mathematician who could find a solution to the many-body problem. This prize eventually went to Henri Poincaré, who invented an array of new analytical techniques for classifying the orbits of three or more objects interacting via gravity. Unfortunately, Poincaré's solution was a negative one—his new techniques showed that, in fact, there *is* no guarantee that a system of many interacting objects will fall into, and continue in, regular orbits.* Poincaré's work is an early landmark in the mathematical study of chaos, and the techniques he invented are still among the standard tools for studying systems that are fundamentally unpredictable despite having relatively simple underlying physics. The long-term stability of the solar system remains in doubt, and thanks to Poincaré we know that this situation can never be resolved.

The situation of many interacting *quantum* particles is even more complicated than the many-body gravitational problem addressed by Poincaré: the electromagnetic force between charges has the same mathematical form as the gravitational force, which already makes stable orbits impossible, and on top of that, it depends on the positions of the interacting particles, which we just showed can't be defined. It's impossible to find a pencil-and-paper solution for the allowed states of even a helium atom, with a single nucleus and two interacting electrons. It's conceivable that some particularly unfavorable arrangement of a large number of nuclei and electrons might be fundamentally unstable. The complex interactions between such a system could end up flinging some particles out to very large distances, while the rest implode to an infinitesimally small point. Which again raises the disturbing possibility of a slice of toast imploding and releasing energy like an atomic bomb.

---

* Interestingly, Poincaré's original conclusion was the opposite—he thought he had shown the ultimate stability of the many-body system. While his manuscript was being prepared for publication, though, one of the journal's editors, the Swedish mathematician Lars Edvard Phragmén, pointed out what seemed to be a small gap in the proof; on closer examination, this turned out to totally reverse the conclusion. The original article had to be hastily retracted and rewritten, but Poincaré still got the prize.

As with Zeno's paradoxes of motion, of course, the ultimate answer is obvious: the fact is, we're surrounded by enormous amounts of matter, in a variety of configurations, and it certainly appears to be stable. Demonstrating this mathematically, however, turns out to be a ferociously difficult problem. It was finally solved in 1967 by Freeman Dyson, who showed that there's a lower limit on the total energy of a collection of electrons and nuclei, ruling out the possibility of implosion. Provided, that is, that the particles involved are subject to the Pauli exclusion principle.

It may not be obvious that the Pauli exclusion principle has anything to do with the energy of confined electrons, but we can see how it works out by looking in more detail at the mathematics of the situation. Pauli's principle is, at a deeper level, a reflection of the fact that electrons are perfectly identical and cannot be distinguished from each other. This means that any labels we place on them for mathematical convenience—calling one A, the next B, and so on, or designating one direction in space as positive and the other negative—are arbitrary. The measurable properties of the many-electron state—including its total energy—cannot change if we swap the labels around. One *un*measurable property can change, though, and in fact is required to: the wavefunction must be "antisymmetric," meaning that when you swap around the labels, it has to change sign from positive to negative. This is the formal mathematical requirement that leads to Pauli exclusion: a wavefunction with two electrons in precisely the same state cannot possibly change sign when you swap the labels, and thus such a state is forbidden.

The sign of the wavefunction doesn't affect the energy directly—the wavefunction, remember, relies on the imaginary number i, so measurable properties can depend only on the *square* of the wavefunction—but this requirement restricts electrons to states that, in general, have a higher energy. We can see how the antisymmetry requirement leads to higher electron energies by considering a simple system that gives rise to two wavefunctions with different symmetries: a single electron shared between two atoms to form a molecule. This isn't precisely identical to the multi-electron scenario, but it's much easier to visualize, and demonstrates why antisymmetric states tend to have higher energies.

A shared electron is attracted to both nuclei, so we expect that a slice through the probability distribution along the axis between the atoms should show two peaks, reflecting an increased chance of finding the electron near each of the nuclei. There are two different ways to make a wavefunction that leads to this sort of probability distribution, though: one where the wavefunction is positive at both peaks, and one where the wavefunction changes from positive to negative as you move from one atom to the other.[*]

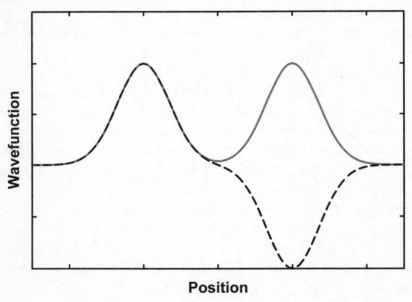

Wavefunctions for the two different states of an electron shared between two atoms.

When thinking about the symmetry properties of these wavefunctions, we need to consider what happens when we swap the arbitrary

---

[*]   As with standing waves on a string, way back in Chapter 2, the actual value of the wavefunction at one of these peaks oscillates through both positive and negative values over time. What matters is the *relative* sign of the two peaks: the same in the symmetric case, opposite in the antisymmetric one.

labels of "left" and "right." This is like reflecting the wavefunctions in a mirror, and we see immediately that the same-sign state is symmetric: both peaks in the wavefunction have the same sign, so if you swap left for right, nothing changes. The different-sign state, on the other hand, is antisymmetric: swapping left for right changes which peak is positive and which negative, which is the same as reversing the sign of the wavefunction.

While there may not seem to be much difference between these, the energy of the antisymmetric state is slightly higher. To understand why, we need to look closely at the probability (shown below) of finding the electron at a point in the vicinity of the molecule—which we get, remember, by squaring the wavefunction, because there can be no negative probabilities.

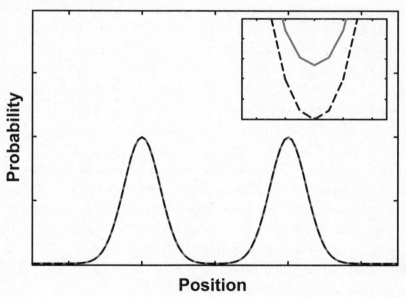

Probability distributions for the two wavefunctions from the previous figure; inset is zoomed in on the midpoint between the two atoms.

These look nearly identical, save for a tiny zone midway between the two atoms. The same-sign (symmetric) state gives the electron some chance of being found exactly halfway between them, while the different-sign (antisymmetric) state has exactly zero probability of being found at the halfway point (since to get from positive to negative, you have to pass through zero). An electron in the antisymmetric state is excluded from a small region of space that an electron in a symmetric state would be free to occupy. That exclusion narrows the range of positions in which an electron might be found, and as we saw when we talked about the uncertainty principle, that necessarily increases the kinetic energy of the particle.

The above illustration involves single-electron wavefunctions, while the Pauli exclusion principle applies to systems with multiple electrons in multiple states, and considers the spin of the electron as well as its spatial distribution (we'll look more closely at this in Chapter 9). The multi-electron problem is much more complicated than the simple one-electron state, but the conclusion carries over: antisymmetric wavefunctions are, in general, slightly higher in energy than symmetric ones, and Pauli exclusion tells us that the wavefunction for a collection of electrons must be antisymmetric. That means that electrons will be found in wavefunctions with a higher energy, so the total kinetic energy of a collection of electrons shared between two atoms increases more rapidly as you pack a number of electrons into a small area than it would if they were not subject to Pauli exclusion.

You can, with considerable mathematical effort, extend this same line of argument to more nuclei and more electrons, and find the same result. A collection of particles subject to the Pauli exclusion principle will always have a higher total energy than an identical number of particles that can occupy symmetric wavefunctions.* And, in fact, this increase in energy is essential to prevent implosion. Just as adding an extra planet can destroy the stability of a solar system, adding additional particles can destroy the stability of a single atom. Without the extra kinetic energy arising from the need to be in antisymmetric

---

* There's a thriving branch of physics dedicated to studying the behavior of such particles, which is essential for understanding superconductivity. Most of the interesting phenomena in this area occur at temperatures within a few degrees of absolute zero, though, so they have little impact on a typical breakfast.

states, a large collection of nuclei and electrons could always reduce its energy to a more negative value by packing more tightly, and solid matter would be inherently unstable.

The mathematical calculation underlying this is formidably complicated, and it wasn't conclusively shown that the energy of a collection of matter had any lower limit at all until Freeman Dyson and Andrew Lenard managed it in 1967, some forty years after the introduction of the Pauli exclusion principle. Dyson and Lenard's work left an uncomfortable amount of room for implosion—their lower limit still would have allowed matter to compress substantially and release tremendous energy, making every solid object a potential nuclear bomb. In subsequent years, Elliot Lieb and Walter Thirring substantially improved Dyson's calculation, and these days we have very solid evidence that solid matter is, in fact, stable just as it is. It's not a surprising result for anyone accustomed to the everyday world, but it is a great comfort to mathematical physicists.

APPLICATIONS TO ASTROPHYSICS

As a postscript to this discussion of the stability of matter, it's interesting to note that when a star dies, the remnant it leaves behind is also held up by the Pauli exclusion principle.

While stars begin their lives with a truly enormous amount of hydrogen, that fuel supply is nonetheless finite and will eventually be exhausted. Once that happens (it can take place in a variety of ways, some more spectacular than others), a core is left behind that can no longer generate energy by fusion. Since the heat released in fusion is what holds an active star up against the attraction of gravity trying to collapse it, this core will shrink inward. As happened during the initial collapse, the energy gained from the inward fall and the electromagnetic repulsion between particles will increase the temperature. Since the 1930s, though, physicists have known that, in the absence of fusion, this increase can't happen fast enough to stop the collapse. The question then is: What happens to the core?

For a smallish collapsing core—up to a mass a bit greater than our sun—Pauli exclusion comes to the rescue. The electrons and nuclei of the core are pulled in by gravity and packed tighter and tighter, to the point where their quantum character comes into play—when the spacing between them becomes comparable to the width of their wavefunctions. Then, just as in the case of solid matter, the fact that they're subject to Pauli exclusion leads to a more rapid increase in kinetic energy than you could get from particles without that requirement. This "electron degeneracy pressure" is enough to withstand the pull of gravity, and the core becomes a white dwarf, an Earth-sized ball of incredibly dense matter held up by quantum mechanics—a one-centimeter cube of white-dwarf matter would weigh several hundred metric tons, compared to a few grams for a piece of rock the same size.

The Pauli exclusion principle alone can't resist gravity for heavier stars, though. Above about 1.4 times the mass of the sun,* the gravitational pull of the core is great enough that the core continues to collapse. Electrons and nuclei are squeezed even tighter, until the distance between them becomes small enough for the weak nuclear interaction to come into play. The weak interaction only works on extremely small length scales, but when matter is dense enough, it lets an electron merge with an up quark, converting a proton into a neutron. In a collapsing core a bit larger than the limit for a white dwarf, the electrons and protons combine to form a mass that consists almost entirely of neutrons.

Neutrons, like protons and electrons, are particles subject to the Pauli exclusion principle. And while they're electrically neutral and thus don't repel each other, the requirement that they be in antisymmetric wavefunctions leads to a rapid increase of energy for sufficiently dense neutrons. This "neutron degeneracy pressure" can halt the collapse of

---

* This is known as the "Chandrasekhar limit" after the Indian American physicist Subrahmanyan Chandrasekhar, who first calculated it on a steamship voyage to England in 1930. Chandrasekhar's initial result met with a good deal of resistance, but he and others repeated and refined the calculations, and he was vindicated mathematically.

a stellar core that's too big to form a white dwarf. This forms a neutron star around ten kilometers in diameter, with a density around a million times that of a white dwarf.

Quantum degeneracy is an amazingly strong force, but in the end, gravity still wins. For a stellar core a bit more than twice the mass of our sun, not even Pauli exclusion can halt the collapse. The neutrons squeeze together more and more, until the whole thing becomes so compact that nothing, not even light, can escape from its surface. At that point, the core forms a black hole, and no further information about its fate is available to the outside universe.

Neutron stars and white dwarfs are some of the most exotic objects in the universe, very far removed from the experience of an ordinary morning. And yet, the quantum properties that keep those extreme astronomical bodies from ultimate collapse are the exact same properties that guarantee the continued existence of you and your breakfast.

CHAPTER 8

COMPUTER CHIPS: THE INTERNET IS FOR SCHRÖDINGER'S CATS

*My tea is still a bit too hot to drink, but I savor the aroma of the rising steam as it cools, and **start up the computer** to see what's going on in the world . . .*

When the Apollo 11 mission landed Neil Armstrong and Buzz Aldrin on the surface of the moon in 1969, it was backed by the best computing power then available. Both the command module, piloted by Michael Collins, and the lander that carried Armstrong and Aldrin boasted state-of-the-art guidance computers. In modern terms, these had about 64 KB of working memory and could carry out around 43,000 operations per second. Mission control back on Earth had five top-of-the-line IBM System/360 Model 75 mainframe computers, each with a megabyte of memory and the ability to perform around 750,000 operations per second.

In the nearly five decades since the moon landing, computing has improved by a mind-boggling amount. I'm typing this book mostly on a Samsung Chromebook with 4 GB of working memory, carrying out some two billion operations per second; the two-year-old smartphone I carry with me runs at a similar speed with slightly less memory. Neither of these is particularly impressive as modern computers go, but they boast thousands of times the processing power of the Apollo program's computers in readily portable packages. Even children's toys these days regularly contain processors far more powerful than those found in the moon lander; if you want to find a processor at the Apollo 11 lander level in a modern device, you probably need to look at a basic kitchen appliance like a toaster oven.

The exponential growth in computing power over the last five decades has been enabled by steady improvements in the manufacturing of silicon-based computer chips. This process requires a detailed understanding of the physics of electrons inside semiconductors, which in turn depends crucially on their wave nature. Ultimately, we have quantum physics to thank for the computers we take for granted nowadays. In fact, the computers we use to share cat pictures over the internet have a deep connection to the most infamous imaginary feline in science, Schrödinger's cat.

THE CAT PARADOX

Famous illustrations of physics concepts, whether thought experiments or actual real-world demonstrations, tend to fall into one of two categories. Many are developed in order to promote a particular new theory by making its successes dramatically visible. While Galileo Galilei most likely did not drop a light object and a heavy one from the Leaning Tower of Pisa (as legend claims), the Flemish physicist Simon Stevin did drop objects of different masses from a church tower in Delft, thus showing that they fall at the same rate. This demonstration has been a staple of introductory physics ever since; the most spectacular variant was done on the surface of the moon in 1971, by Apollo 15 astronaut Dave Scott.

New theories are also often promoted with thought experiments, such as the "light clock" introduced by Gilbert Lewis and Richard Tolman in 1909 to explain some of the central ideas of Einstein's special relativity. They imagined an unusual sort of clock that marks time by bouncing light back and forth between two mirrors, recording a "tick" at each reflection.* An observer with such a clock watching an identical one move past will see the light in the moving clock follow a longer path than the light in the stationary clock, and thus take a longer time between "ticks." This elegantly uses the constancy of the speed of light to explain one of the central features of special relativity, namely that moving clocks run more slowly than stationary ones, and it also explains how the effect is relative—a second observer traveling with the moving clock will see *that* clock tick at its normal rate, while the first observer's clock runs slow.†

The other category of famous physics illustrations are puzzles intended to show subtle problems in reasoning when applying a particular theory. The famous "twin paradox" of special relativity is one such thought experiment. It imagines sending one of a set of twins on a long rocket voyage, while the other remains home on Earth. The prediction that moving clocks run slow suggests that the twin on the rocket should experience a shorter time than the twin left on Earth, and return to find her sibling dramatically aged. Then again, though, the relativity of motion suggests that the twin in the rocket should see her sibling as the one who is "moving," which suggests that the Earthbound twin should be younger. Seemingly solid reasoning leads to the conclusion that each twin must be younger than the other, a paradoxical result.

Of course, physical reality can't accommodate a true paradox, so only one of the two twins can be "younger" than the other. The apparent paradox is resolved by noting that the twin in the rocket necessarily

---

* The ticking of any reasonably sized light clock would be impractically rapid, but in principle it would be a great clock as the constant speed of light would give it a very regular oscillation.

† For a much longer discussion of this, and the rest of Einstein's most famous theory, see *How to Teach Relativity to Your Dog* (Basic Books, 2012).

changes their speed and direction of motion—accelerates—making their situation distinguishable from their sibling's, allowing for a definite difference in elapsed time. A small-scale version of the experiment has even been done with atomic clocks on jet airplanes, and the results match the predictions from relativity: the clock on the plane runs slower than its twin left on the ground, by the expected amount.

The puzzle of "Schrödinger's cat" falls into the latter category—paradoxes that illustrate problems with the reasoning of an existing theory. In 1935, both Erwin Schrödinger and Albert Einstein had become dissatisfied with the quantum theory they had played crucial roles in launching, and each wrote papers describing thought experiments to illustrate their view that quantum theory was fundamentally incomplete and needed to be replaced with a deeper, more satisfying approach to physics. (Einstein's paper, written with young colleagues Boris Podolsky and Nathan Rosen, introduced the problem of what we now call "quantum entanglement," which will be the subject of Chapter 11.)

Einstein's hypothetical demonstrated that indeterminate quantum states are incompatible with the principle of "locality"—the idea that the state of an object should only depend on things in its immediate vicinity. Schrödinger's contribution to the world of quantum puzzles was in a similar vein: like Einstein, he was bothered by the probabilistic nature of quantum mechanics, and how it is that the single reality we observe emerges from a sea of possible outcomes. In the view of quantum physics promoted by Bohr and others—known as the "Copenhagen interpretation" after the location of Bohr's institute—this issue was sort of swept under the rug by asserting that the probabilistic rules of quantum physics only applied to microscopic systems, and they could not affect the macroscopic world. Schrödinger didn't buy this, and he invented a diabolical thought experiment to highlight the issue.

In a paper summarizing "The Present Situation in Quantum Mechanics," Schrödinger pointed out that you could connect microscopic quantum physics to macroscopic effects in a dramatic way. He imagined the scenario of a cat sealed into a box with a device containing an unstable atom with a 50 percent chance of decaying from one of its allowed states to another in the next hour (as in Einstein's statistical photon model and Heisenberg's matrix mechanics). If the atom decays, the device would then immediately poison the cat. The box is sealed in such a way that the experimenter outside has no way of knowing what's happened until it's opened one hour later. The question is: What is the state of the cat just before the box is opened?

Common sense would seem to say that the cat is either alive or dead, but according to the Copenhagen picture, the state of the *atom* must be indeterminate: an equal mix of decayed and not decayed, up until the box is opened and the final state determined. Mathematically, the wavefunction for the atom contains two pieces, one corresponding to each of the possible states, in the same way that the wave packets we put together in the previous chapter contain multiple possible momentum components. The state of the atom is a quantum superposition: an indeterminate state that is not definitely either, but both at once.

The connection between the atom and the cat-killing machine, though, makes the state of the *cat* wholly dependent on the state of the atom, so the cat must be in a quantum superposition as well, both alive and dead *at the same time*. Schrödinger's thought experiment shows that the Copenhagen interpretation's reliance on an absolute separation between the microscopic world of atoms (governed by quantum rules) and the macroscopic world (where classical physics holds sway) simply isn't feasible. The two can be connected, as demonstrated by the cat puzzle, which forces physics to grapple with the underlying issues: How is the single reality we see picked out from quantum probabilities? What does it mean to measure the state of a quantum object? And what does it mean for a quantum object to exist in multiple states at once?

The problem of the cat in a box helped spark a conversation about fundamental quantum issues that continues to this day. It has also inspired numerous experiments to create "Schrödinger cat states" where a quantum object is placed in a superposition of two distinct states.* Nobody has done (or would do) this with a literal cat, but "cat

---

* There isn't universal agreement among physicists on what counts as a "cat state." The term is often used for states where single quantum particles are

states" have been produced in a wide range of systems—single atoms, ions, large numbers of electrons inside superconductors—and there's an active field of experimental physics working toward making cat states in ever larger objects.

These experiments are extraordinarily difficult, and they generally require elaborate equipment employed in carefully controlled laboratory conditions. The underlying physics principle, however—that quantum objects can exist in superpositions of multiple states—is well established. In fact, it is essential for understanding the behavior of any number of everyday objects, from simple molecules on up to computer chips.

CHEMICAL BONDS AS CAT STATES

The electron pair-bonding paradigm has been essential for understanding chemistry since even before quantum mechanics. The notion of bonds formed by sharing electrons to produce filled shells is still an essential part of chemistry, but the development of the full theory of quantum mechanics gives new insights into what that actually means.

In the days of the Bohr-Sommerfeld model, there were some attempts made to explain molecular bonds in terms of defined electron orbits around both nuclei in a diatomic molecule—large ellipses and figure-eight loops, and so on. These were never terribly successful, and when the discovery of matrix mechanics and the Schrödinger equation destroyed the idea of well-defined electron orbits in atoms, it became obvious that this picture was clearly inappropriate. Modern quantum chemistry, like modern atomic physics, describes the electron in terms of a spread-out wavefunction.

We looked at these wavefunctions briefly when we discussed the stability of matter in the last chapter. As you may remember, a

---

placed in a superposition of two states, but others argue that it should properly refer only to superpositions involving an object made of a macroscopic number of particles. This state of affairs is probably fitting for an analogy involving cats.

one-dimensional slice through the wavefunction of an electron in a diatomic molecule looks something like the following illustration.

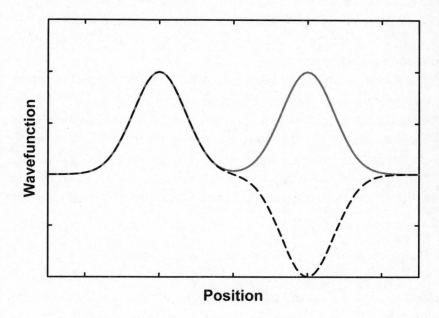

There is a peak in the wavefunction (and thus the probability of finding the electron in that area) near the location of each nucleus, and the electron spreads out over a wider region of space to encompass both atoms.

The wider spread of the electron in the two-atom molecule helps explain why molecules form in the first place. An electron with a larger range in space will tend to have a lower energy than one that is more tightly confined, as we saw in Chapter 7. The exact details will depend on the specifics of the atoms involved, but in general, pairs of atoms form chemical bonds because spreading the electrons across both nuclei lets them reduce the total energy of the pair.

Looking more closely, if we compare this molecular wavefunction to the wavefunctions of the electron for each individual atom, an interesting insight emerges: the wavefunction for the electron in the molecule resembles the sum of the wavefunctions for electrons bound to two individual atoms. This idea is the starting point of many techniques for calculating the states of electrons in molecules.* In the same way that we put together wave packets by adding wavelengths when we discussed the uncertainty principle, you can build up wavefunctions for shared electrons by starting with the states of single atoms and combining them to find an accurate representation of the electron wavefunction for a molecule.

In this view, an electron in a molecule is in the same sort of superposition state as Schrödinger's imaginary cat. The electron is not bound to atom A or atom B, but to both A *and* B at the same time. This gives us another way of thinking about what it means for electrons to be "shared" in the shell-filling model of chemistry, and it's also a tool for understanding what happens as you start to consider the quantum properties of solids containing uncountably large numbers of atoms.

MORE IS DIFFERENT

When we move from talking about single molecules to objects large enough to see, we run into exactly the problem that the Copenhagen interpretation was trying to duck: macroscopic objects don't appear to be very quantum. Where single atoms absorb and emit light in narrow, discrete spectral lines, macroscopic solids tend to absorb and emit light over broad ranges of wavelengths. The crystals used as the gain medium in some lasers, for example, can emit light at wavelengths spanning several hundred nanometers in the red and near-infrared regions of the spectrum. Lasers made with these crystals can be tuned to any wavelength in that span by using a filter to select a particular wavelength to be amplified.

In the same way that the narrow spectral lines emitted by atoms suggest the discrete energy levels of Bohr's atomic model, the broad

---

*   The wavefunction is not perfectly identical to the sum of two individual atomic states, but it's close. There are well-honed mathematical techniques for starting with the sum-of-two-atoms wavefunction and tweaking it slightly to better match the actual molecular state.

emission of solids and large molecules suggests that in these systems, electrons can take on a wider range of energies. Further evidence of this is found in the electrical behavior of materials: a small voltage applied to a piece of conducting material will cause electrons to flow through it readily; the current increases smoothly with the voltage, with no sign of abrupt jumps between quantum states. In a macroscopic material, then, the electrons seem to be able to take on any velocity they like, unlike the discrete states of atoms. In fact, you can do a remarkably good job of describing the electrical properties of metals with a simple model in which the electrons move freely through the material, only occasionally bouncing off an atomic nucleus.

This is an example of the phenomenon—well-known throughout the sciences—of "emergence," famously explored by the Nobel laureate Philip Anderson in a 1972 paper titled "More Is Different." As Anderson pointed out, there are numerous situations in which a sufficiently large collection of simple objects—atoms, molecules, cells—that interact with each other by one set of rules will exhibit collective behaviors that are described by a completely different set of higher-level rules.

As Anderson noted, this is responsible for a certain hierarchical structure to the sciences—biology is just chemistry applied to a sufficiently large collection of molecules, and chemistry is just physics applied to a sufficiently large collection of atoms, and so on. It also allows for an enormous richness of real-world phenomena and approaches to studying them, because the higher-level rules are not necessarily related to the more fundamental ones in an obvious way.

But while the higher-level rules may not be suggested obviously by the more fundamental ones, that doesn't mean they're not connected—the high-level rules must indeed emerge from the fundamental. We know from innumerable experiments on single atoms and photons that physics at the microscopic level is thoroughly quantum, so the behavior we see in macroscopic objects must be able to be explained in terms of the quantum rules for simple systems applied to large assemblages of atoms. In a sense, the problem of electrons in a macroscopic material is just an illustration of one of the chief dilemmas facing quantum physics, the one Schrödinger was pointing at with his infamous zombie cat: How do we get to the classical rules that govern the world we see from the principles of quantum physics?

There are hints, even in macroscopic objects, of some underlying quantum behavior, particularly when it comes to their electrical properties: solids absorb and emit light in broad ranges of frequencies, but those ranges don't cover the entire spectrum. There's a minimum wavelength to the light absorbed or emitted by a typical material, and that's a clue. It's also true that while current flowing through a conductor behaves like free electrons moving with no quantum jumps, many other materials are insulators, with electrons that seem to be locked in place, unable to move without a significant input of energy.

So, our project for the rest of this chapter is to show how those classical-ish properties (broad emission and absorption spectra, and a lack of quantum jumps in the current flowing through a conductor) emerge from the quantum rules governing electrons and atoms. This will also explain those hints of quantum behavior (gaps in the spectrum and the properties of insulators) that are visible in macroscopic materials. Most importantly, a thorough understanding of the quantum underpinnings of electrical properties reveals ways to control and manipulate those properties, and as we will see, this knowledge is what enables us to create the materials that allow the construction of computer chips.

FROM SPECTRAL LINES TO ENERGY BANDS

We can begin to understand the shift from discrete allowed energy levels to broader energy bands by thinking about what happens as a single electron is shared between more and more atoms. As we've said, the wavefunction of a shared electron in a molecule amounts to the sum of the wavefunctions for the individual atoms involved. And if we start with an electron shared by a single pair of nuclei, as we saw in the last chapter, there are two possible wavefunctions—we used these to illustrate the difference between symmetric and antisymmetric states. In "cat state" terms, we can think of these as "left plus right" (the symmetric case) and "left minus right" (antisymmetric). These give rise to very similar probability distributions, but the "left minus right" state involves a small excluded region midway between the two atoms, and thus has a slightly higher energy. When we bring two atoms together, then, what was a single well-defined energy state for each separate atom splits into two energy states in the molecule. One state shifts up slightly, and the other down a similar amount (because the electron is spread over a wider area), so the range of possible energies for an electron has increased.

If we bring in a third atom, we add additional possibilities. In terms of wavefunctions, we get states that look like this:

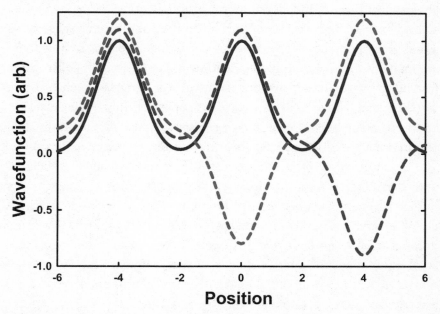

Wavefunctions for an electron in a three-atom molecule. These are slightly offset from one another vertically to make them easier to see.

In the cat-state language, these are "left plus middle plus right," "left plus middle minus right," and "left minus middle plus right." The all-plus state, like the symmetric state for a pair of atoms, has no excluded regions—points where the wavefunction goes to zero—

meaning it has the lowest energy of the three. The "left plus middle minus right" state has a single zero point, increasing its energy slightly, while the "left minus middle plus right" state has *two* excluded regions (because it changes from positive to negative and then back), shifting its energy up even more.

Thus, the single energy state for an electron in a single atom has increased to three states for an electron in the molecule, with three different energies. As we move from one atom to two to three, the range of possible energies for the electron increases, and this process continues as we add more and more atoms: each new atom in the chain gives a new set of states with different numbers of zeroes, and thus different energies.

For molecules with a smallish number of atoms, this turns the single spectral lines of individual atoms into collections of many closely spaced lines. Light is still absorbed and emitted at discrete wavelengths as electrons jump between states of well-defined energy, but now there are many of these close together, leading to larger numbers of possible transitions with similar but not identical wavelengths. Each of these states can be thought of as a cat-like superposition of the electron bound to many individual atoms at the same time, with the individual atomic wavefunctions positive for some atoms and negative for others.

As the number of atoms increases, the spectral lines associated with these states begin to blur together. By the time you have a few million atoms, let alone the 10^{23} atoms it takes to make up a visible chunk of solid matter, it's no longer feasible to talk in terms of a finite number of discrete energy states. Instead, we speak of electrons in a solid occupying continuous bands of energy, with gaps between them. Absorption or emission of light then involves moving an electron from a particular energy within one band to a particular energy within another.* The change in energy, and thus the wavelength and frequency of the photon involved, can take on many different values. An electron with an energy near the bottom of a low band might absorb a photon

---

* Electrons can also move *within* bands in response to an electromagnetic field, as we'll see when we talk about conductors and insulators, but this doesn't involve the absorption or emission of visible light.

with a very short wavelength and move to a state with an energy near the high end of the upper band. On the other hand, an electron with an energy near the bottom of the upper band dropping down to an energy near the top of the lower band will emit a photon with a rather long wavelength. Solids interact with light in the same general ways as atoms—absorption, stimulated emission, and spontaneous emission*—but they have a wider range of options when it comes to the wavelength of the light involved.

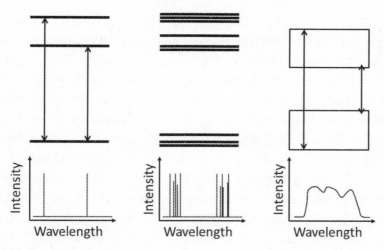

Development of energy bands. Left: An atom with discrete energy levels, and widely spaced spectral lines. Center: A small molecule has energy levels split, leading to many more closely spaced lines. Right: A macroscopic solid has nearly continuous energy bands, allowing for broad ranges of absorption and emission.

Calculating exactly which wavelengths can be absorbed or emitted is a complicated process, and the result depends on the exact three-dimensional arrangement of atoms within the solid; enormous amounts of computational effort go into working out the details of

---

*   There are additional processes as well, involving interactions with the crystal lattice, which lead to many interesting complications that keep solid-state physicists happily occupied. These processes don't figure into any of the phenomena we'll talk about, though.

the band structure of real solids. Thinking of electrons as Schröding-er's cats shared among many different atoms, though, gives us a way to understand the essential phenomena: as the number of atoms increases, energy states multiply and smear out into bands, with energy gaps between them that determine maximum and minimum values of the wavelength ranges that can be absorbed and emitted.

WHY BAND GAPS?

The argument above explaining the origin of energy bands may leave you wondering why there should be gaps between the energy bands at all. If every new atom added to the structure of a solid slightly increases the spread of possible electron energies, it might seem that the energy bands should broaden until they merge together, leaving electrons free to take any energy they like. This doesn't happen, though—even in the largest solids, there are ranges of energy that are completely forbidden. These "band gaps" exist because of the wave nature of electrons, and a version of the same phenomenon produces the vibrant colors of tropical birds.

One of the more amazing interactions between physics and biology shows up in some species of parrot—specifically in their brilliant blue feathers, which do not contain any blue pigment. That is, if you exactly matched the chemical composition of a blue feather, and made a solid block of that material, it would not appear blue. In chemical terms, the feathers are made of the same protein as human fingernails, which by itself is sort of grayish and translucent. The color is not intrinsic to the material, but arises from the internal structure of the feathers.

If you use an electron microscope to look at the blue feathers of a tropical bird, you will find a spongy network of filaments of keratin with gaps of a few hundred nanometers between them. These gaps, combined with the wave nature of light, produce the blue color we see by preventing blue light from traveling through the material.

We can understand the way this works by considering a one-dimensional slice of this material, with light that can only travel straight forward or straight back impinging on a regular array of filaments spaced by a few hundred nanometers. As the light wave travels along, each time it encounters a filament, it reflects a tiny amount of light straight back.

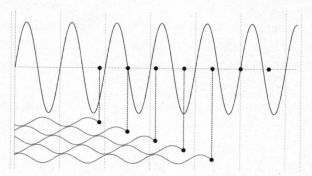

Waves encountering filaments in a material when the filaments are closer together than a wavelength. The reflected waves end up out of phase with each other and will interfere destructively, leading to no reflection.

Each of these reflected waves adds together with the incoming wave and all the other reflected waves from other filaments. If the spacing of the filaments is small compared to the wavelength of the light, this leads to reflected waves with lots of different phases, and when you add them all together to determine the total amount of reflected light, they mostly cancel each other out. Very little light of that color reflects, and thus the wave passes on through the material with only a small attenuation.

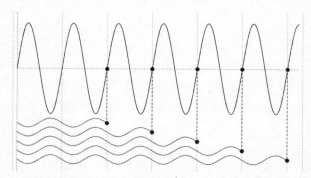

Waves encountering filaments in a material when the spacing matches the wavelength. The reflected waves all start with the same phase, and add constructively with each other and destructively with the incoming wave, preventing the light from traveling within the material; thus, almost all of the light is reflected.

When the spacing between filaments closely matches the wavelength of the incoming light, though, each reflected wave ends up in phase with all the others and out of phase with the incoming light. When you add the reflected waves in this case, you find that they all combine to produce a larger reflected wave, which cancels out the incoming wave. Light at that wavelength is thus unable to travel through the filament network and instead gets reflected.

The spongy network of filaments making up tropical bird feathers have spaces between them of around 400 to 500 nanometers in size, comparable to wavelengths at the blue/violet end of the visible spectrum. Light in the red part of the spectrum, with a wavelength of 600 to 700 nm, passes through, but blue light is strongly reflected, giving the feathers a brilliant blue color without any blue pigment.*

Light at wavelengths significantly shorter than the size of the spaces between filaments should also be transmitted—provided that the spacing is not an integer multiple of the wavelength, in which case the various reflected waves would again be in phase with one another, giving reflections at shorter wavelengths. These aren't relevant for determining the color of the feathers, though, because the wavelengths involved are much too short for human vision.

We can understand the band gaps in solid materials in similar terms. The atoms in a solid material form a crystal lattice—a regular array of atoms spaced by the length of the molecular bond between the atoms (this is typically around 0.2 nm, but it varies a bit depending on the particular elements involved and the type of bond). As electron waves move through this lattice, they will scatter off the atoms making up the lattice, sending waves back the way they came. At energies where the electron wavelength matches the spacing between atoms in the lattice,

* This structural color phenomenon turns up in many species of birds, but generally only for shades of blue. Red feathers on tropical birds take their color from red pigment molecules, so the material of the feather itself is colored. The wave nature of light is also exploited by birds and butterflies to make iridescent colors that appear to shift when you change the viewing angle; this is a different process involving interference off different layers of a structure like thin overlapping scales.

these reflected waves add together and cancel out the original wave, meaning that electrons with those energies simply cannot exist inside the material. This guarantees that no matter how many atoms you have in the lattice, the energy bands will always be separated by gaps where the wavelengths of electrons at that energy line up too neatly with the spacing between atoms.

These two effects, the cat-state–like sharing of electrons between all the different atoms in a solid, and the wave interference that produces band gaps, provide the basis for our modern understanding of electrons inside matter. When you take both of these effects into account, for a sufficiently large number of atoms, you end up with a set of broad allowed energy bands, spaced by band gaps whose energy and width depends on the arrangements of atoms in the crystal.* That structure of bands and gaps, combined with Pauli exclusion, not only explains the electrical properties of most ordinary matter, it is what lets us manipulate silicon to make modern computing possible.

INSULATORS AND CONDUCTORS

Thinking about the sharing of electrons between atoms and the motion of electrons within a crystal lattice explains how the narrow allowed states of atoms become, in the molecules making up a solid, broad energy bands separated by gaps. What remains to be explained by our quantum picture is how this determines a material's electrical properties. It turns out to be a bit like chemistry: in the same way that the

---

*   We are, of course, skipping lightly past a number of technical details—working out the correct band structure of a three-dimensional crystal is a computationally intensive process, and it consumes a great deal of time on the part of physicists generally; measuring these band structures experimentally to test the calculations is another important source of activity. What we've described here is just the conceptual underpinnings of a large and active field of research. And, of course, there's a bit of spherical-cow modeling going on here as well, because not all materials have a nice, regular crystal structure; dealing with substances that aren't regular crystals is another important research area.

chemical reactivity of an element is determined by how the electrons fill up the available states in an atom (atoms with only partially filled outer "shells" will more readily react by giving up or receiving electrons), whether a given material is an insulator or a conductor will likewise depend on how the electrons fill up the energy bands in the solid. The electrical properties ultimately depend on where the energy of the last electron put into the solid falls within the band structure.

At first glance, determining "the energy of the last electron" may seem like an impossible proposition, since a continuous band of energies would involve an infinite number of possible states within a tiny range. Thinking of bands as continuous is only a matter of convenience, though—in reality, the bands are still made up of discrete states of well-defined energy; there are just so many states so close together that they *look* like a continuum. But there are, in fact, a finite number of states, so as we imagine adding electrons into the bands, Pauli exclusion tells us that each electron fills up a particular state, forcing the next electron to go elsewhere.

The first electron goes into the lowest energy state available, filling it up, so the second electron goes into the second-lowest energy state, and so on, in just the same way that electrons going into atoms fill up electron shells leading to the different chemical properties of the elements in the periodic table. There's a bit more math involved in the process than with an atom, given the nearly infinite numbers of states and electrons involved, but there are well-understood tools from calculus for dealing with these sorts of issues. Both the number of states and the number of electrons available to fill them increase as we add more atoms, but those two effects balance each other out, and in the end, we find that for a given substance with a particular crystal structure, the electrons end up filling all the states up to a particular energy.

The energy of the last electron added to the sample is called the "Fermi energy" after Enrico Fermi, who developed the statistical techniques needed for describing the states of large numbers of electrons. This energy can be fairly substantial—if we think of it as the kinetic energy of the moving electron, it corresponds to around a million meters per second, or a temperature of tens of thousands of degrees.

For these states, which involve electrons shared through the whole material, that's not a picture that should be taken too literally—a given electron is not in a particular place zipping through the material at almost 1 percent the speed of light—but it gets the scale across: there's a very large difference in energy between the first and last electrons put into a solid. This energy is associated with the motion of the electrons within the solid in the same way that an electron inside an atom has some kinetic energy, but it isn't literally orbiting in the way envisioned in the original Bohr model.

This picture of high-energy electrons makes understanding the flow of current a little more complicated, but it's not as bad as you might think. The Fermi energy defines the base state of a material with nothing else going on: the electrons are moving around with their characteristic internal energy, but as a whole, they're not going anywhere. Loosely speaking, at any given time there are as many electrons moving to the left as to the right, so there's no net movement of electrons from one place to another. For all the frantic motion it implies, an unperturbed quantum solid with electron states filled up to the Fermi level behaves pretty much like a classical one in which no electrons are moving at all.

A conductor with an electric current passing through it, on the other hand, involves the flow of electrons in a particular direction. In the picture of energy bands, this means that, for example, some of the electrons that are initially moving to the right must move to the left instead, to give a net leftward flow of electrons.* This can't come from simply redirecting electrons with energies below the Fermi energy, though, because all the leftward-moving states below the Fermi energy are already filled, by definition. To create a net movement of electrons to the left necessarily requires moving some electrons to states with energies above the Fermi energy.

---

* Which is described as a "conventional current" flowing to the right, a quirk of physics that has been confusing students of electronics for generations. You can blame Ben Franklin for this—he was an influential proponent of the modern model where one type of charge moves and the other remains stationary. Unfortunately, he guessed wrong when assigning a positive value to the mobile charges.

If the Fermi energy falls somewhere in the middle of a band of allowed energies, this is a relatively simple process because there are empty states just above the Fermi energy. The additional energy needed to excite an electron up to an open leftward-moving state is minimal, and it's easily provided by applying a small voltage. The difference between energies is so small, we don't see it as a quantum jump—it looks like a smooth increase in the energy from a state where nothing is moving to a state with a small number of electrons moving in a particular direction. Materials with Fermi energies in partially filled bands are thus electrical conductors.

On the other hand, if the Fermi energy lies at the top of a filled band, the next available state in which electrons could move in the appropriate direction is on the far side of the band gap. This requires a much larger input of energy to get an electric current flowing, generally comparable to one short-wavelength photon per electron excited. That might make such a material useful as a detector of light, with current flowing through it only when light shines on it and excites some electrons, but the necessary energy isn't easy to get by applying a voltage, and it very definitely looks like a quantum jump. Materials with Fermi energies at the top of a band are thus electrical insulators, and will not carry electric current except under extreme circumstances.

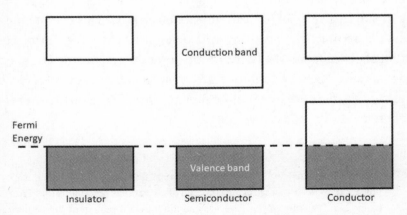

Energy bands for insulators, conductors, and semiconductors. States below the Fermi energy are filled with electrons (shaded region).

SEMICONDUCTORS AND THEIR USES

For most everyday purposes, "conductors" and "insulators" are the most important classes of materials when dealing with electricity. Insulators are materials like wood, plastic, and rubber that protect you from electric currents, while conductors are mostly metals, and terrible things to stick in a wall socket. We can use our quantum model of electrons in a solid to understand how these categories arise, and how to sort different materials into them.

The real test of the power of a scientific model, though, is not just its ability to explain simple and obvious phenomena, but the ability to predict other, more subtle effects suggested by the model's underlying principles. The best models allow scientists to exploit these underlying phenomena to make new and useful things, and for this reason, the most important application of quantum physics to solids is in the area of semiconductor materials.

Semiconductors are, as the name suggests, not especially good conductors on their own. Their band structure, however, allows their conductivity to be manipulated by small changes in their composition, and this is the feature that provides the final link between cat states, Pauli exclusion, and the computer chips that turn up in everything these days.

In band structure terms, a semiconductor is just an insulator with a relatively narrow band gap. The Fermi energy lies at the top of a full energy band, but the energy gap between the full "valence" band and the empty "conduction" band is small enough that heat energy within the sample can naturally excite some electrons. As with Planck's oscillators way back in Chapter 2, each electron gets a share of the thermal energy in the material. The average energy given to any one electron is small compared to the band gap, let alone the Fermi energy, but a few electrons can receive way more than the average energy and end up in the higher band. This puts a few electrons into states where they can readily conduct, because there are plenty of empty states corresponding to movement in whatever direction you like, so the material is capable of carrying a small electric current.

The elements silicon and germanium are examples of natural semi-conductors, but pure samples of these materials are not especially interesting or useful. What makes them useful is that a tiny admixture of something else can dramatically increase the conductivity, by one of two different means.

One way to increase the conductivity of pure silicon is to add a very small amount of an element from the next column to the right in the periodic table—typically phosphorous or arsenic. These elements have one additional electron, but otherwise are chemically similar to silicon in many respects, so they fit into the lattice in a way that doesn't perturb the band structure too much, provided the amount added is small—typical values work out to around one phosphorous atom per million silicon atoms. This is why silicon computer chips are manu-factured in "clean room" environments by people wearing full-body spacesuits: a tiny level of contamination by outside particles during the manufacturing process can mess up the whole process. The primary change this "doping" makes is to add some discrete states with elec-trons at energies just below the conduction band. The extra electrons that start in these states are very readily excited to the conduction band, where they increase the semiconductor's ability to carry electric current.

It may seem like adding electrons to the conduction band would be the only way to increase the conductivity of a semiconductor, but in fact the opposite process also works. Doping pure silicon with elements from the column to the *left* in the periodic table will also increase the conductivity, by removing electrons from the valence band. An atom like boron also has a strong resemblance to silicon, chemically speak-ing, but with one electron fewer. A tiny admixture of boron happens to add a few *empty* states with energies just above the valence band, into which electrons from the lower band are easily excited, and once there, they get stuck.

It may not seem like removing electrons from the valence band would increase the conductivity, but it does, in an interesting way. Trapping those silicon electrons on the dopant boron atoms leaves "holes" in the sea of electrons filling the material. When a voltage is

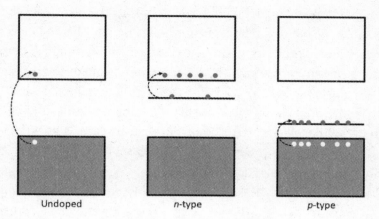

Band diagrams showing different types of semiconductors. In an undoped semiconductor, thermal energy excites a tiny number of electrons from the valence band to the conduction band. In an n-type semiconductor, a donor level just below the conduction band supplies many more electrons, increasing the conductivity. In a p-type semiconductor, an acceptor level just above the valence band traps some electrons, leaving behind holes that can carry current.

applied to the material to drive a current, the remaining electrons will shift around in response, which changes the position of the holes in such a way that they seem to move in the opposite direction from the electrons.

These gaps in an otherwise full band of electrons thus behave as if they were positively charged particles moving in an otherwise empty band. The motion of these holes carries current in a very similar way to the motion of the electrons in a metal—or a phosphorous-doped sample of silicon—leading to a higher conductivity for the material.[*]

So, both adding and removing electrons from a semiconductor can boost its conductivity. There are a few critical differences between "n-type" semiconductors (involving adding electrons, as in phosphorous-doped silicon) and "p-type" semiconductors (in which electrons are removed, as in boron-doped silicon), mainly involving

_____

[*]   Unlike electrons, though, the "holes" move in the same direction as the conventional current through the material.

their behavior in a magnetic field, which pushes the positively charged holes in the opposite direction from the negatively charged electrons. This phenomenon allows a simple experiment to distinguish between the two when investigating the properties of new materials. This magnetic response is also the basis for the magnetic field sensor in your smartphone, which allows it to act like a compass when you're navigating in unfamiliar places. But aside from that, whether a given piece of doped semiconductor is p-type or n-type doesn't much matter.

Something amazing happens, though, if you stick an n-type semiconductor onto a p-type semiconductor of the same base composition (for instance, both silicon). When you apply a voltage across the junction between these materials, the difference between the types of charge carriers leads to dramatic differences in behavior depending on the sign of the voltage applied to each side. If you apply a positive voltage on the p-type material and a negative voltage on the n-type, current will flow. The holes in the p-type material move away from the positive voltage, toward the boundary between the two, and the electrons in the n-type material move away from the negative voltage, also toward the boundary. When the two meet, the electrons flowing to the boundary from the n-type material fill in the holes flowing in from the p-type material. Meanwhile, new electrons are pushed into the n-type material at the negative-voltage end, while electrons are pulled out at the positive-voltage end, creating new holes. This process can continue indefinitely, and current will readily flow through the junction.

If you reverse the voltages, however, the situation is very different. A negative voltage applied to the p-type material will draw the positive holes to it, away from the boundary, while a positive voltage on the n-type material draws electrons to it. This produces a very brief current as the material rearranges itself, but in the absence of a source of new electrons, the current can't be sustained.

So, while doped semiconductors by themselves are not especially interesting, a junction between a p-type and an n-type semiconductor material creates something very interesting indeed. The combination of the two makes a diode, a device that will only allow current to flow in one direction. This finds all sorts of applications in everyday

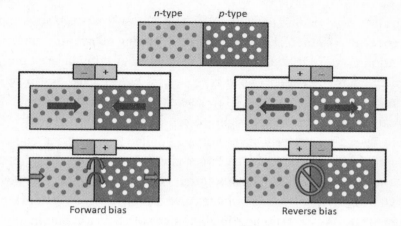

n-type   p-type

Forward bias   Reverse bias

The motion of electrons and holes in a diode for different applied voltages. A negative voltage on the n-type semiconductor pushes electrons toward the boundary, where they combine with holes pushed away from the positive voltage on the p-type, allowing a continuous flow of current as new electrons flow into the n-type and out of the p-type. Reversing the voltage pulls electrons toward the positive voltage and holes toward the negative, leaving a depleted region at the boundary and stopping the flow of current.

technology, mostly protecting components that can only tolerate current flowing in one particular direction. And with the right choice of semiconductor materials, the electrons recombining with holes at the junction between materials will release a photon whose frequency is determined by the band gap of the semiconductor. Such light-emitting diodes (LEDs) have been used for low-energy lights on clocks and other devices for decades. More recent improvements in LED technology* have made them essential components of computer displays and residential lights. LEDs can also be used as the basis for a type of laser, using the polished front and back faces of the semiconductor chip as the "mirrors" for the laser cavity (as described in Chapter 5). The result is a powerful laser source in a package only about a centimeter across, and these are used for reading and writing data on

---

*   The 2014 Nobel Prize in Physics was awarded to Isamu Akasaki, Hiroshi Amano, and Shuji Nakamura for the development of LEDs that emit blue light.

optical storage media (like CD, DVD, and Blu-ray players), supermarket checkout scanners, and laser pointers, among many other things.

Adding a third layer of material—sandwiching a thin layer of p-type semiconductor between two n-type layers, say—makes a still more interesting device. This three-layer stack looks a bit like two diodes back to back, and with the right choice of doping levels for the different layers, a relatively small voltage applied between one end and the middle can trigger a much larger current to flow from the other end through the middle. The amount of current that flows tracks with the voltage—larger applied voltages give more current. This device is a transistor, a key component in all manner of electrical amplifiers—a "transistor radio," the cutting-edge technology of the 1950s, is one that uses compact transistors to amplify electric currents to drive the speakers, instead of the bulky and hot vacuum tubes used for previous radios. This enabled the first readily portable, battery-powered audio players, setting the stage for the Walkman, the iPod, and eventually the ubiquitous smartphones of the modern era.

If you design your electronics to use only two voltage levels, rather than the continuously varying level of an audio signal, a transistor will function as a digital switch—current either flows or it doesn't—and this is the crucial element for computer processors. A whole array of transistors can be used to represent numbers in binary form, and more complicated circuits of transistors can perform mathematical operations on those numbers.

This is the basis for modern computing technology. The first general-purpose electronic computers were built in the 1940s, and were based on large numbers of vacuum tubes. Not long after the invention of the first transistor in 1947,* semiconductor-based transistors began to replace vacuum tubes, first as stand-alone components and then in "integrated circuits," where multiple electronic components are built

---

* The physicists John Bardeen, Walter Brattain, and William Shockley won the 1956 Nobel Prize for their invention, one of two Nobels shared by Bardeen; the other was in 1972 for a theory of superconductivity developed with Leon Cooper and Robert Schrieffer.

into a single block of silicon. This is done by varying the doping of different layers of the material so that they are arranged to make transistors, then etching this material into transistors that are a few nanometers on a side.

A single chip a centimeter or so square can contain billions of interconnected transistors, arranged into the circuits needed to process binary data. These semiconductor "chips" are much more compact and require less electrical power than vacuum tubes, and they quickly became the standard for electronic data processing.

The Apollo Guidance Computer referred to at the start of this chapter was one of the earliest integrated-circuit computers,[*] and since then, the performance of chip-based computers has improved exponentially, to the point where a slightly-out-of-date smartphone puts many times the processing power needed to land men on the moon into a device that fits readily into a pocket.

All that semiconductor-based processing power, along with the LEDs that provide your screen's display and the high-power transistors that amplify the sound, is made possible through quantum mechanics. Understanding how the wave nature of electrons leads to the band-gap structure in large collections of atoms—and how that structure can be manipulated to change the electrical properties of a material—is essential to the design of not only our laptops and desktop machines, but of the computers found in nearly everything these days, from refrigerators to cars to toasters. The modern picture of electrons as waves whose behavior is governed by Schrödinger's famous equation, and the tendency of those waves to spread themselves among multiple states at once in the same manner as his infamous cat, is what ultimately allows us to turn otherwise boring chunks of silicon into revolutionary technology.

---

[*]   Though this computer was a bit of a hybrid, as many of its instructions were hardwired into "core rope memory" consisting of small coils of wire.

CHAPTER 9

MAGNETS: HOW THE H*CK DO THEY WORK?

*I open the refrigerator to start breakfast, careful not to dislodge the many works of art **held to the door with magnets** . . .*

The force between two magnets, or between a magnet and a piece of metal, is among the most captivating examples of fundamental physics—for young and old. One of the most enduringly popular toys at my kids' day care is a set of plastic tiles in simple shapes that snap together thanks to magnets in the edges; almost every day, these are built into elaborate new structures. The local science museum gets a lot of mileage out of a giant horseshoe magnet and several handfuls of steel washers, and adults are as likely as kids to be found trying to see how long a chain of washers they can stick to one of the poles.

In fact, magnets are a great gateway drug to a career in physics. Einstein recalled being captivated by a compass as a child, his wonder at the invisible force that always pulled the needle back to the north launching a lifetime of speculation about the forces of nature. Most physicists I know have childhood memories of, for example, trying to get a small magnet to levitate over a collection of larger ones.* Even for adults, the fascination remains, and magnetic desk toys are a common feature of faculty offices in physics departments everywhere.

As familiar as they are, the working of magnets is also famously difficult to explain. A frequently shared interview clip from the 1980s shows renowned physicist Richard Feynman declaring flatly that "I really can't do a good job, any job, of explaining magnetic force in terms of something else you're more familiar with, because I don't understand it in terms of anything else that you're more familiar with."† A less high-brow example is the 2009 song "Miracles," in which the rapping-clown group Insane Clown Posse triggered a thousand unsuccessful attempts to explain magnets with the line "F*cking magnets, how do they work?"

It may seem strange that a phenomenon so common that we use it to hold stick-figure drawings to kitchen appliances is so hard to describe in nontechnical language, but the physics is, in fact, extremely compli-cated, and depends on subtle details of the microscopic structure of particular materials. And, of course—as you probably guessed—it ulti-mately traces back to the quantum: the permanent magnets we use to hold pieces of paper up for display would be impossible without elec-tron spin and the Pauli exclusion principle.

NAVIGATING MAGNETISM

When people ask "How do magnets work?" they are really asking two separate but related questions. A permanent magnet is a macroscopic chunk of material that produces a magnetic field in its vicinity, and one

---

* This won't work with stationary magnets, but if you make the magnet part of a rapidly spinning top you can, in fact, get it to hang in midair. A toy version of this is available under the name "Levitron" and makes a useful physics demonstration.

† This is a little unfair to Feynman, who was, in fact, making a larger point about the problem of "why" questions in general. It's regularly used as a disclaimer before attempts to explain the physics of magnetism, though, so it clearly resonates on that level.

way of interpreting the question is in reference to the general behavior of these magnetic fields. This is, from the standpoint of physics, the easier of the two questions to tackle. The nature of magnetic fields has been understood since the mid-1800s, when Maxwell wrote down his equations showing how currents and changing electric fields create magnetic fields, and vice versa.

Unfortunately, while Maxwell's equations offer a straightforward way to understand how magnetic fields are created by moving charged particles around, they don't answer the other question about permanent magnets, namely why those specific chunks of otherwise inert material spontaneously generate magnetic fields in the first place. After all, there don't seem to be any currents flowing in a hunk of naturally occurring magnetite, and yet this mineral produces a significant magnetic field. The tendency of certain minerals to attract metals has been known since at least the sixth century BCE, recorded in Greece and India and China, and it's been put to practical use since at least the eleventh century CE, by which time the Chinese were using magnetic compasses for navigation. Despite that long history, though, the origin of the magnetic properties of these minerals remained a mystery into the twentieth century.

The existence of permanent magnets defies easy explanation because it involves physics on many levels. Physics on the scale of atoms is obviously involved, because naturally occurring magnetic materials all contain iron, and only a few other elements are clearly magnetic. Atomic-scale physics is not the whole story, though: many materials containing large amounts of iron are *not* magnetic, including many steel alloys, so the crystal structure of the material must also play a role. And, of course, everything is ultimately pieced together from fundamental particles, so magnetic behavior must have roots in the behavior of individual protons and electrons.

The most useful application of magnets, the constancy of a compass's direction, also highlights the other issue that adds complexity to the problem of magnetism: magnetic interactions are fundamentally more complicated than the electrostatic attraction or repulsion between charged particles. The electric charge of a particle is a single value, and if you know the charge, you immediately know the force on that particle due to an electric field. The energy of two interacting charges depends on the sign and size of their individual charges, the distance between them, and nothing else.

There is no magnetic analogue to a single electric charge, however—you never find a magnetic north pole alone without a matching south pole—so magnetic forces depend not only on a simple charge but also on a direction. As anyone who has played with bar magnets knows, the force between two magnets gets stronger or weaker, and even changes from attractive to repulsive, depending on which direction the north pole of each is pointing. To find the energy of a pair of magnets, you need to know not only their strength and separation, but also the angle between their north poles.

This dependence on orientation adds some additional overhead when trying to determine the behavior of particles with magnetic character. Like the electric field, the magnetic field comes with an associated direction, but determining its effect on a magnetic particle placed in the field also requires keeping track of a direction associated with the particle. That extra information also adds to the bookkeeping required to calculate the properties of a large collection of magnets, and opens the possibility of entirely new collective phenomena. A large collection of magnets all pointing in the same direction is a very different thing than a collection where each magnet has its north pole in the opposite direction from its neighbor's.

We can cut through some of the complexity involved in the multiple scales of magnetism by using the same fundamental principle that explains so much of physics: no matter what scale we look at, any physical system is always trying to find the lowest energy state possible. Finding that minimum energy involves balancing the energy costs of all the different interactions that a given object—whether it's a fundamental particle, an atom, or a small chunk of mineral—has with the rest of the universe. Keeping that balancing act in mind provides a simple and reliable guide to navigating the complexity of permanent magnets, like a compass needle always pointing the way north.

In general, the energy of a magnetic object, at whatever scale, will be lowest when its north pole is pointing in the same direction as the magnetic field at its position, and highest when it points in the opposite direction. This is what makes a compass work: currents in the core of the earth generate a magnetic field on a grand scale, so that every point on the surface of the planet sits in a small magnetic field pointing in a particular direction. A compass needle is a small, light, permanent magnet that's able to freely rotate about its center to minimize its energy, which happens when the north pole of the magnet is pointed toward the North Pole of the planet, more or less. We designate the poles of magnets as "north" or "south" depending upon which geographic direction they point to when allowed to rotate freely. By convention, though, the field surrounding a magnetic object points outward in the region of the magnet's north pole and inward in the region of its south pole, with the field lines in between forming closed loops, like those traced by the familiar demonstration of scattering iron filings over a bar magnet. This north-to-south direction of magnetic fields means that what we call the earth's "North Pole" actually corresponds to the south pole of a typical magnet.*

Aligning individual magnets with the field produced by other nearby objects not only changes the energy of the magnets, but also how their individual fields add up to produce the field around the group. If the magnets are positioned end to end, the lowest-energy arrangement will have all the north poles pointing in the same direction; in this case, their individual magnetic fields add to make a stronger effective magnet. On the other hand, magnets placed side by side will prefer their north poles to be in opposite directions, in which case their individual fields will largely cancel out, making a weaker effective magnet.

---

* This is a great quiz question in introductory physics classes. The north magnetic pole is also slightly offset from the north end of the axis about which Earth rotates, so depending on your position, magnetic north deviates slightly from true north; the difference between the two is well-known, however, and marked on good navigational maps.

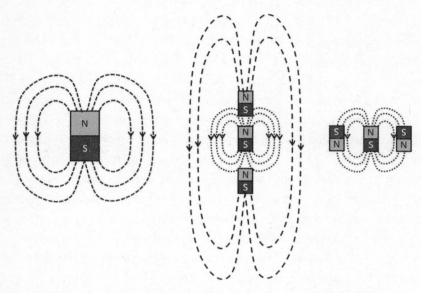

The magnetic field lines for a single magnet, and the lowest-energy configurations for groups of multiple magnets. End-to-end magnets with their north poles aligned generate a larger collective magnetic field, shown by the larger loops, while the fields of side-by-side magnets cancel each other out.

A three-dimensional material made up of smaller particles with magnetic character will necessarily have some of those particles placed side by side, which is why the vast majority of materials are nonmagnetic. Even strongly magnetic atoms like iron and chromium end up in nonmagnetic forms when combined into minerals or alloys, because the lowest-energy way for those magnetic atoms to arrange themselves in molecules and crystals has the north poles of neighboring atoms pointing in opposite directions.

Making a strong permanent magnet requires finding a way to put particles together so that the minimum energy at every scale—that of fundamental particles, magnetic atoms, and chunks of mineral—comes when the north poles of the individual magnetic components are aligned. This can't be managed with magnetic interactions alone; it requires an additional interaction that increases the energy of the non-magnetic state so that the magnetic state is preferred. This is very tricky to arrange, and in the end requires us to factor in not only the electrostatic repulsion between electrons, but also, once again, the Pauli exclusion principle.

MAGNETIC ELECTRONS

Magnetism begins at the level of fundamental particles, and the intrinsic magnetic character of electrons is the ultimate source of the magnetic field of a permanent magnet. The interaction between pairs of elementary particles also provides a clear illustration of the energy balancing that governs the whole process.

As we saw when we first introduced the notion of Pauli exclusion in Chapter 6, a single electron has "spin," a purely quantum property that can take on only two possible values. This spin gives the electron a small amount of magnetic character, and in the presence of a magnetic field, the two values of spin produce two states of slightly different energy. These states are traditionally called "up" and "down," depending on whether the electron's internal magnet points in the same direction as the local magnetic field or in the opposite direction.

Of course, the magnetic character of the electron doesn't just give it a preferred direction, it also *creates* a magnetic field, which affects other nearby particles. A second electron placed in the vicinity of the first will tend to align its spin with this field, giving that second electron a preferred direction that depends on whether it's end to end or side by side with the first. If we considered magnetic interactions only, the electrons would tend to arrange themselves into long chains with neighboring chains having alternating spin, the whole arrangement in the end producing no net magnetic field.

Of course, two electrons in close proximity don't interact only via their magnetic properties; they also feel electrostatic interactions, and they repel each other very strongly because they have the same negative charge. This repulsion is vastly stronger than the tiny magnetic interaction, so two electrons don't stick around long enough for the magnetic interaction between their spins to matter. While the pair of electrons can lower their energy by pointing their spins in opposite directions, they can lower the energy *much more* by moving farther apart, and as a result, they end up separated by enough distance that the tiny magnetic interaction makes no discernible difference.

This magnetic character does have measurable effects, though, when two particles with spin can be induced to hang around each other a little longer. If we take an electron and a positron—the positively charged antimatter version of an electron—and bring them close together, they can form a short-lived "atom" held together by the attraction between their charges. As in an ordinary atom, the two particles can lower their energy by drawing closer together, but putting them into a smaller volume causes an increase in their kinetic energy, and the balance between these two determines the atom's optimum size. Their mutual attraction keeps the electron and positron in this "positronium atom" close enough together that their magnetic interaction produces a measurable effect. The lowest energy state for positronium is split into two states depending on the relative alignment of the spins of the electron and positron: when both north poles are in the same direction, the energy is slightly higher, and when they're in opposite directions, the energy decreases. The "hyperfine splitting" between these states has been measured experimentally: positronium has a spectral line in the microwave region of the spectrum, corresponding to photons with a frequency of about 203 GHz.

This magnetic interaction also comes into play in more ordinary matter. A proton *also* has a quantum-mechanical spin, and thus produces a magnetic field, so an electron bound with a proton to make a hydrogen atom also has its energy shifted by the magnetic interaction between them, splitting hydrogen's lowest energy state into two. The energy separation corresponds to a photon with a frequency of 1.4 GHz, in the radio region of the spectrum,* and light emitted by hydrogen moving between these states is one of the principal tools used by radio astronomers to study distant clouds of hydrogen gas.

---

*   The shift is much smaller in hydrogen than positronium because the magnetic
      field generated by a proton is much smaller than that of an electron or positron.

The magnetic interaction energy in both of these cases is only a tiny perturbation to the electrostatic interaction—the energy difference between the two hyperfine levels in positronium is about 1/10,000th of the energy difference between the two lowest-energy electron orbitals. This is why the original Bohr model was able to completely neglect magnetic interactions: at the scale of fundamental particles, the electrostatic interaction absolutely dwarfs any magnetic effect. As we move to the scale of multi-electron atoms, though, the situation becomes more complicated, and as the Pauli exclusion principle comes into play, the extreme strength of electrostatic interactions becomes a crucial factor in producing magnetic atoms and minerals.

MAGNETIC ATOMS

One tempting but wrong idea about the origin of magnetism at the scale of atoms is that it is the result of orbiting electrons behaving like the current in an electromagnet. While this would fit nicely with Maxwell's equations of classical electromagnetism, it doesn't fit the evidence. Every atom in the universe consists of electrons orbiting a nucleus, but only a handful of elements in the middle part of the periodic table show significant magnetic character. Magnetism in atoms can't be solely a result of electron orbits.[*]

The idea of orbital motion as a source of magnetism was behind the original Stern-Gerlach experiment, discussed back in Chapter 6, in which a beam of silver atoms was split by a special magnet. Unfortunately, as the physicists who grappled with Stern and Gerlach's results found, that theory didn't match the behavior of the atoms—differences in orbital motion ought to split the beam into at least three components, where Stern and Gerlach saw only two. Their result helped point toward the existence of an electron property with only two values,

---

[*] Somewhat loosely speaking, this is because an electron is as likely to be orbiting clockwise as counterclockwise, and the magnetic contributions from those two possible orbits cancel each other out.

namely spin; for our current purposes, it's also a clear hint that magnetism in atoms is ultimately due to the spin of their electrons.*

Making a magnetic atom, then, is a matter of getting the tiny magnetic fields produced by the electrons inside the atom to add together
to make a bigger magnet. This means getting the electron spins pointing in the same direction, so their "north poles" align. This goal faces a
major obstacle, though: the fact that the magnetic interaction between
electrons favors states where the spins point in *opposite* directions.

At first glance, the Pauli exclusion principle, which forbids any two
electrons from having exactly the same quantum state as determined
by the four quantum numbers n, l, m, and s, would seem to make this
worse, because it builds in this kind of pairing of electrons. As we saw
in Chapter 6, the lowest energy state for the electrons in any particular
atom is found by "filling up" the available energy states of the atom
(determined by n, l, and m) with at most two electrons each: one spin
up ($s = +\frac{1}{2}$), the other spin down ($s = -\frac{1}{2}$). This natural pairing of spin
up and spin down explains why none of the atoms near the edges of the
periodic table are strongly magnetic. Those elements have their outermost energy levels nearly or completely filled, with their electrons
paired up so their magnetic fields cancel out.

In elements near the middle of the periodic table, though, Pauli
exclusion combines with the repulsion between electrons to create a situation where the electron spins *want* to line up with each other. This
has to do with the deeper meaning of the Pauli principle discussed in
Chapter 7, as a requirement on the symmetry of a collection of electrons.

An element from the middle few columns of the periodic table will
have its outer shell half full of electrons, which seems to give it several
options for how to arrange those electrons and their spins. The canon-
ical magnetic element, iron, for example, has six electrons to place in a

---

*   The orbital motion of electrons does affect their interaction with magnetic
     fields, leading to the Zeeman effect, where a single energy state splits into
     multiple sublevels when an atom is placed in a magnetic field. These sublevels
     do not create a magnetic field outside the atom that could be used to power a
     permanent magnet, though.

state with $l = 2$, which has five distinct sublevels of the same energy but different values of m. There are lots of ways to arrange these electrons, but for the purposes of understanding iron's magnetic properties, we can focus on only two: one where all six electrons are clustered in just three of the sublevels, and another where the electrons are spread more evenly, with only one sublevel having an electron pair.

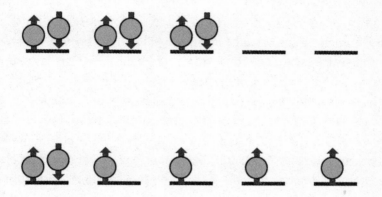

Two possible arrangements of electron spins for the half-filled outer shell of iron, one nonmagnetic (top), the other magnetic.

The Pauli exclusion principle dictates that whenever two electrons are paired up in the same sublevel, they will have opposite spins. Both of these states satisfy Pauli exclusion, but the one with all six electrons paired up is nonmagnetic, while the more distributed state has four unpaired electrons pointing in the same direction, giving it strong magnetic character. The energy of all five n, l, and m sublevels is the same in both arrangements, however, so it may seem like there's no reason one should be any more likely than the other.

That analysis, though, neglects the energy contributed by the repulsive interaction between nearby electrons. This increases as the separation between electrons decreases, and a pair of electrons occupying the same spatial sublevel would be very close together indeed. The repulsion between paired electrons raises the energy of the nonmagnetic state, making the magnetic state with aligned spins the lowest-energy state available.

You might reasonably object that you could make a nonmagnetic state with the electrons distributed over more sublevels, by flipping the spins of two of the unpaired electrons, so that the state has one electron pair, two single spin-up electrons, and two spin-down electrons. But the symmetry aspect of the Pauli exclusion principle takes care of that; how it does so is easiest to understand if we consider only two electrons and two sublevels.

As discussed in Chapter 7, the Pauli exclusion principle states that the wavefunction for a multi-electron state must be antisymmetric. Because electrons are identical and interchangeable, the measurable properties of the state as a whole cannot change if we swap the labels on two electrons—but the combined wavefunction must change sign after the swap. This antisymmetry requirement applies to the wavefunction as a whole, both the spatial distribution of electrons (determined by n, l, and m) and the distribution of their spins, which means that if one of these is antisymmetric, the other must be symmetric. If both spin and space wavefunctions were antisymmetric, a swap of labels would change the sign twice, putting you right back where you started—in physics as in English, two negatives (awkwardly) make a positive.

Thus, if the two spins point in the same direction, the spin wavefunction is symmetric, and the space wavefunction must be an antisymmetric combination of the two available sublevels. If the spins point in opposite directions, that can be an antisymmetric state,* in which case the space wavefunction must be symmetric.

We've seen that, for a space wavefunction, antisymmetric states exclude the electrons from more space, and that raises their energy slightly, which might make you think that these would be the higher-energy states—and for a single electron, the antisymmetric state is indeed higher energy. The antisymmetric arrangement keeps the electrons farther apart on average, though—you can get a sense of why

---

* There's also a symmetric combination of one spin-up and one spin-down electron, which is often grouped together with the both-up and both-down states, collectively referred to as a "triplet" state in contrast to the "singlet" antisymmetric state.

by thinking about the two-atom states we looked at back in Chapter 7. The excluded region for those wavefunctions is the spot midway between the two atoms, which pushes the two peaks a tiny bit farther apart.

The antisymmetric wavefunctions for electrons in a single multi-electron atom are not split between positions around two nuclei like those in molecular states, but rather are superpositions of different n, l, and m states around a single nucleus. The end result is the same, though: the electrons in an antisymmetric combination of orbitals are a tiny bit farther apart, on average, than those in a symmetric combination. That increase in distance reduces the energy due to their mutual repulsion by more than the energy difference between symmetric and antisymmetric spatial wavefunctions.

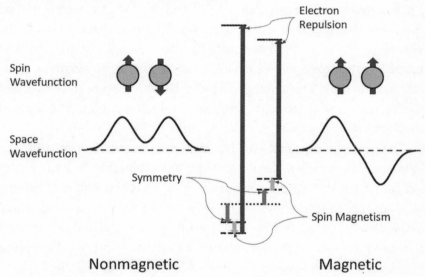

The process that makes the magnetic arrangement favored in multi-electron atoms. The nonmagnetic arrangement features a symmetric spatial wavefunction and a favorable alignment of spins, both of which lower the energy compared to a state without those effects included (dotted line), but the repulsion between electrons in this state is very strong. In the magnetic arrangement, the antisymmetric spatial wavefunction and the magnetic interaction between spins both slightly increase the energy, but the reduction in the repulsive interaction between electrons is more than enough to compensate.

Thus, the lowest-energy state available to iron is one in which the outer-shell electrons are distributed among all the available sublevels, with the spins of unpaired electrons aligned. This means that the magnetic fields created by the individual spins add together to produce a larger field, making iron a strongly magnetic atom. The same basic physics is at work in other elements with half-filled outer shells, leading to the cluster of atoms with strong magnetic character in the middle columns of the periodic table.

MAGNETIC CRYSTALS

Of course, as noted above, just because an *atom* of a particular element is magnetic doesn't mean that a solid chunk of that material will be a permanent magnet—if it did, naturally occurring magnets would be everywhere. In fact, some elements that are strongly magnetic at the atomic level (chromium, for example) show almost no magnetic character at all in bulk. The making of a permanent magnet requires not just aligning the spins of electrons within an atom, but aligning the spins of atoms within a crystal.

The phenomenon that makes a magnetic mineral is ultimately the same one that makes a magnetic atom: a combination of Pauli exclusion and repulsive forces that goes by the (somewhat misleading) name "exchange interaction." The structure of a crystal is determined by the sharing of electrons, which establishes the distance between atoms and their three-dimensional arrangement. This crystal structure then determines the energy bands and band gaps for the electrons in the material, as we saw in Chapter 8, which in turn determines many of their electrical properties.[*]

When we talked about molecules and solids in previous chapters, we mostly ignored the effect of spin (other than the state-filling effect of Pauli exclusion) and interactions between electrons, but just as they

---

[*]   This may seem a little circular, with the electron states determining the arrangement of atoms and then the arrangement of atoms determining the

do at the atomic level, these play a key role in magnetism at the level of macroscopic materials. The calculations become much more complicated to carry out, but the mutual repulsion between electrons still increases the energy of states where the electrons are close together. This repulsion tends to be smaller for antisymmetric spatial states, and when electrons are in antisymmetric spatial states, their spins are lined up.

For the right combination of materials, the iron atoms in a mineral end up separated by just the right distance so that their total energy is lower when the electrons in the crystal fall into antisymmetric space wavefunctions. This means that the spin wavefunctions must be symmetric, with their spins pointed in the same direction and adding together to make a stronger combined magnet.

Getting just the right distance between magnetic atoms depends on subtle details of the chemistry and crystal structure, which is why magnetic minerals are so rare. Even alloys made entirely of magnetic elements can be made nonmagnetic by changing the mix of atoms. A stainless steel alloy consisting of mostly iron with about 15 percent chromium will naturally be magnetic. On the other hand, a different alloy that increases the chromium slightly and adds a bit of nickel (around 8 percent) is nonmagnetic.

This magnetic behavior is also very fragile—the energy shifts involved are generally quite small, and depend again on subtle details of the crystal structure. Some nonmagnetic alloys can even be made magnetic solely by mechanical manipulation: the stainless steel alloy typically used for kitchen appliances is technically nonmagnetic, but the process by which the panels are shaped deforms the crystal structure somewhat, which is why we can use magnets to stick crayon drawings to our "nonmagnetic" stainless steel refrigerators.

---

electron states. Theoretical calculations of these things usually involve an iterative process: picking a plausible arrangement of atoms, then calculating the electron states, then recalculating the arrangement of atoms to see if the new electron states favor a shift. In nature, this process just happens automatically; it's much easier to be an atom than a theoretical physicist.

When all the various factors involved come together in the right way, the electrons in a particular region will tend to align their spins with those of their nearest neighbors, making a small "magnetic domain" of that piece of the crystal, which acts like a microscopic magnet. Even this is not enough to make a permanent magnet, though. Naturally occurring chunks of metal consist of enormous numbers of little crystals with slightly different orientations, each making a domain with its north pole pointing in a random direction.

If a magnetic material consisting of many little domains pointing in random directions is exposed to a strong magnetic field—say by placing a magnet next to the surface—each of those domains can lower its energy by shifting its electrons around to align with the field. This produces a large number of domains with their south poles pointed at the north pole of the magnet, and it's responsible for the attractive force between a magnet and a piece of metal. This alignment of domains is only temporary—when the magnetic field is removed, the individual domains return to their original random orientations.

Making a permanent magnet requires rearranging these domains in a more lasting way. This can be done mechanically—if you're patient, you can turn a steel paper clip into a weak permanent magnet by rubbing it with another magnet—or by heating the material to a high temperature and letting it cool in the presence of a strong magnetic field.* This results in a material where the electrons in all the individual domains have their spins (more or less) aligned in the same direction, adding together to make a stronger magnet.

Once established, a permanent magnet, as the name suggests, will tend to keep this alignment, even though the crystal structure of the individual domains might favor a different arrangement. While the material's

---

* Or even a relatively weak one—rocks cooling in the magnetic field of the Earth become slightly magnetized as a result. This is one of the decisive bits of evidence for continental drift: on either side of the mid-Atlantic ridge, we see a pattern of "stripes" with alternating magnetization, as the Earth's magnetic poles have reversed direction many times over millions of years. New rocks formed as magma moves up and out through the ridge trace the history of pole shifts and the spreading of the ocean floor.

total energy could be lowered by having the electrons point in the right direction for each domain, the energy would have to *increase* in the intermediate steps of this process. Again, though, this magnetism is easily disrupted: as a material is heated, the thermal energy added to the motion of the electrons can become large enough to cover the energy increase needed so that electrons will be free to orient their spin however they like—usually in the direction favored by the crystal structure of their particular domain. Magnetic materials thus have a characteristic "Curie temperature" above which their electrons will no longer remain aligned across different domains, and they lose their magnetic character.[*]

Understanding the physics involved, from electron spins up to crystal domains, has also allowed physicists to engineer magnetic materials that are not found in nature. In particular, since the 1970s, the use of extremely strong magnets based on "rare earth" elements like neodymium has become widespread—they're found in everything from kids' toys to magnetic data storage systems. These have made magnetic fasteners in general much more common, and more reliable than they were when I was of an age to make drawings to stick on the refrigerator.

MAGNETIC DATA STORAGE

While the realignment of domains in a magnetic material placed in a magnetic field is usually temporary, for some materials, applying a sufficiently large field can force a more permanent realignment. Once aligned, these domains will remain in their new orientation after the field is removed, until something else—heating, mechanical manipulation, or a strong enough field in a different direction—disrupts the new arrangement. This persistence of magnetic domains has made these materials an essential part of the data storage industry.

---

[*]   This is named after Pierre Curie, whose original research was in the physics of magnetic materials. As Marie Curie began to work on radioactivity, though, Pierre abandoned magnets to join her in those experiments, which we'll discuss in the next chapter.

In the early days of computers, many machines used "magnetic core memory," where bits being used in computation were stored temporarily in small chunks of magnetic material, with the direction of the north pole switched between two values by running a current through a loop of wire around each bit. The magnets in these could be fairly substantial—large enough to create signals picked up by a nearby radio. One of my computer science professors in college told a story about designing a punch-card program that would pointlessly flip bits in the right pattern to play the Sesame Street song "Rubber Duckie" on a radio left next to one of these computers.

On a smaller scale, flexible strips of magnetic material formed the basis for the cassette and VHS tapes that were staples of my teenage years, storing sounds and video in patterns of magnetic domains written onto the tape using electromagnets in the recorder. These patterns were then read out by a small detector picking up the changing magnetic field caused by the tape passing beneath a coil of wire in the player. Tapes could store data for long periods of time, though the materials used would slowly degrade after many playbacks.

In less obsolete technologies, rewritable magnetic domains are also behind the operation of modern hard disks. The basic principle remains the same: an electromagnet in the "write head" changes the orientation of magnetic domains on the disk to store digital information. Meanwhile, the "read head" detects the pattern of magnetic fields on the disk, converting the stored information back into ones and zeroes in working memory. Decades of engineering effort in developing better magnetic materials and high-performance data-writing and -reading systems has pushed these drives to the point where they can store an incredible amount of data. The four-terabyte drive I use to back up my computer at home is about the size of a box of the 5.25″ floppy disks used by my first computer; that whole box would've held around one millionth of the data of my current backup drive.

This chapter has only skimmed lightly over the extremely complex physics of magnetic materials, a rich and varied field keeping huge numbers of physicists happily occupied. Whether you're inter-ested in high-density data storage or just displaying crayon drawings on kitchen appliances, though, all of this physics is deeply rooted in quantum mechanics. Every magnet you encounter is ultimately a quantum object, drawing on the intrinsic spin of the electrons within it.

CHAPTER 10

SMOKE DETECTOR: MR. GAMOW'S ESCAPE

It's still dark in the hallway when I leave the bedroom, **the status light on the smoke detector** *casting a faint light on the wall.*

When I was in graduate school in the middle to late 1990s, I lived in Rockville, Maryland, where I rented a room in a house that had the strangest smoke detector I've ever encountered, in that it went off nearly every time I made toast. I didn't have to burn the toast, mind you—the mere act of toasting bread would somehow trigger the alarm, which tolerated all manner of other cooking, and also one housemate who smoked multiple packs of cigarettes a day.

Many years later, I'm still totally at a loss to understand what it was about toast, specifically, that set off this smoke detector. While an explanation for that behavior remains out of reach, though, the basic operation of a *normal* smoke detector is fairly straightforward. It's also dependent on another of the famous oddities of quantum physics—the ability of particles to pass through barriers that classical physics says should stop them cold.

THE CLASSICAL PHYSICS
OF SMOKE DETECTION

Smoke is, pretty much by definition, a collection of small particles lofted into the air by a flame. Detecting smoke, then, means detecting these particles rapidly enough to alert homeowners to a fire before it can harm them.

The simplest way for a device to detect smoke is essentially the same way we perceive it with our eyes: looking for the scattering of light by smoke particles in the air. Smoke becomes visible to us either by reflecting light that otherwise wouldn't have reached our eyes, or by blocking light that otherwise would have. A photoelectric smoke detector relies on the former: a small light source shines through a tube, with a light sensor placed off to the side. Under ordinary conditions, no light hits the sensor, indicating that everything is fine. When smoke particles enter the tube, some of the light bounces off to the side, generating an electronic signal from the light sensor, which triggers an ear-splitting beep.

Certain kinds of fast-burning fires can produce particles that don't scatter much light, though, and another detector technology uses radioactive decay to pick these up. In an ionization detector, a stream of alpha particles is sent into a small air chamber between two charged metal plates. When an alpha particle strikes an air molecule, the collision can split the molecule into two charged pieces, one positive and one negative. The positive ion is drawn toward the negative plate of the detector, and the negative ion to the positive plate, and the arrival of these particles leads to a small flow of current through a circuit containing the plates.

In the absence of any smoke particles, the flow of current is fairly constant and produces the "all is well" signal for the device. When smoke enters the ionization chamber, though, the smoke particles absorb some of the ions and prevent them from reaching the plates, disrupting the current's flow. This drop in current is registered by the electronics in the detector, and triggers the ear-splitting beep.

These two different detector technologies each have advantages and disadvantages, and as a result many commercial smoke detectors use both in parallel. Each also relies to some extent on quantum physics. The first type detects light through the photoelectric effect, which (as we talked about back in Chapter 3) was ultimately explained by the existence of photons. For the second type of detector, the quantum connection is more direct, and it comes from the ionization process, which relies on alpha particles generated by the decay of an artificial radioactive element, americium-241, placed within the detector. This decay involves a mystery that predates quantum physics, one that was eventually solved by a colorful character from the USSR.

THE MYSTERIES OF RADIOACTIVITY

In the late 1800s, physics was rocked by the discovery of two seemingly new forms of radiation. First, in 1895, Wilhelm Conrad Röntgen stumbled upon x-rays while experimenting with the effects of electric current flowing through vacuum tubes. Röntgen noticed that even after he had enclosed his apparatus to block the escape of light, a fluorescent screen across the lab would glow faintly when current was flowing in the tube. He correctly attributed this to some extremely penetrating rays emanating from the device, and in short order had produced a now-iconic x-ray photograph of his wife's hand, clearly showing the bones. His work almost immediately found medical applications, and in 1901 he was awarded the very first Nobel Prize in Physics for his discovery.

As surprising as x-rays were, Röntgen's vacuum-tube apparatus was at least doing something to actively supply the energy needed to generate radiation, by passing an electrical current through the tube. When the current was shut off, the production of x-rays ceased.* The next discovery was far more puzzling: Henri Becquerel, following up on

---

* Today we know that the x-rays are produced when electrons passing through the vacuum tube strike the positive electrode at high speed.

Röntgen's work, found that uranium compounds emit x-rays and other radiation *all the time*, with no energy input at all. This seemed to involve the spontaneous creation of energy from nowhere—which, according to accepted laws of physics, is impossible—and launched an effort to identify the sources of radioactivity.

One of the most successful scientists to investigate radioactivity (and in fact the coiner of the term "radioactivity") was Marie Skłodowska Curie, who began experimenting on uranium compounds soon after Becquerel's announcement, and showed that the radiation originated within uranium atoms, not as the result of some chemical process involving interactions within a larger molecule. She also discovered that some ores containing uranium were even more radioactive than the uranium refined from them, indicating the presence of some other, unknown radioactive element.

Marie Curie embarked on a long project to identify and isolate this new element, and eventually her husband Pierre joined her. Working together in a makeshift lab in a courtyard at the University of Paris that the German chemist Wilhelm Ostwald described as "a cross between a stable and a potato shed," the Curies discovered two new elements, polonium* and radium, leading to two Nobel Prizes. In 1903, the Curies and Becquerel shared the Physics prize† for their experiments on radioactivity, and in 1911, Marie alone‡ won the Chemistry prize for isolating radium and polonium.

At around the same time, Ernest Rutherford, then at McGill University in Montreal, was conducting his own experiments on radioactivity, and he developed the modern classification of radiation into alpha, beta, and gamma forms. These were ordered in terms of their penetrating power, with alpha particles the least penetrating (alpha emission is easily blocked by a few sheets of paper), and gamma rays the most

---

*   Polonium was named in honor of Marie's home country of Poland, then part of the Russian empire.

†   Initially the Royal Swedish Academy of Sciences had planned to give the prize to the two men only, but after Pierre Curie objected, they recognized Marie as well.

‡   This likely would have been shared with Pierre, but he was killed in a traffic accident in 1906 and the Nobel Prize is not awarded posthumously.

(penetrating some distance even through dense materials like lead). In 1900, Becquerel showed that beta particles are high-energy electrons, and in 1905 Rutherford found that alpha particles are doubly ionized helium; gamma rays were shown to be high-energy photons in 1914.

Radioactivity was a fertile area of research in the early 1900s, as both a subject of study in its own right and a tool for investigating other questions. The 1909 Marsden and Geiger experiment in Rutherford's lab that revealed the existence of the nucleus (discussed in Chapter 4) was carried out using the high-energy alpha particles emitted by radium. What process produced that radiation, though, and particularly where the necessary energy came from, remained a mystery.

The problem is demonstrated most clearly by measurements made by Hans Geiger in 1921, when looking at the interactions of alpha particles with uranium. Shooting high-energy particles at uranium atoms showed that the repulsive interaction between the uranium nucleus and positively charged alpha particles would push away particles with an energy of about 8.6 MeV (million electron volts) or lower,* which is consistent with what you would expect for the charge of a uranium nucleus. Uranium itself is radioactive, though, and emits alpha particles with an energy of about 4.2 MeV—much lower than the minimum energy needed to get an alpha particle *into* the nucleus.

If we look at the problem in energy terms, it's clear why this is impossible in classical physics. A particle has two kinds of energy: kinetic energy due to its motion and potential energy due to its interactions, whether repulsive or attractive.

The strong nuclear interaction is a powerful attraction but acts over a short range, making the alpha particle's potential energy negative only at very small distances from the nucleus. At extremely long distances, the strong force doesn't matter at all, and the electromagnetic repulsion between the nucleus and the alpha particle is still tiny.

---

* Geiger was limited to naturally occurring radioactive sources, which mostly produce lower-energy alpha particles, so he wasn't able to measure an exact limit by shooting in particles at energies too high to be deflected, only to infer a minimum value from measurements at lower energy.

In the intermediate range, the electromagnetic repulsion between the two positively charged particles is significant, but the strong interaction has not yet kicked in. So, if we start out at long distances and move toward the nucleus, we see the alpha particle's potential energy slowly rise from zero to some peak value, then dive down to a large negative value once it is close enough to the nucleus to feel the strong attractive force.

Putting all this together, the potential energy of our alpha particle passing close to the nucleus of some atom (moving from right to left in the figure below) looks like this:

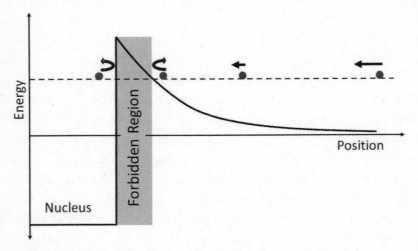

Energy diagram for an alpha particle near a nucleus. Long-range electrostatic repulsion combined with the strong nuclear interaction produces a potential energy barrier that traps alpha particles inside the nucleus and turns back alpha particles coming in from the right.

How does this limit the motion of an alpha particle? We can see this by remembering that the total energy of a particle—kinetic plus potential energy—must remain constant. An alpha particle starting a long way outside the nucleus and moving at some speed will have a positive total energy due to its motion, and basically zero additional energy due to interactions, as it is too far away to feel either repulsion or attraction from the nucleus. As it approaches the nucleus, the potential energy increases as it begins to feel electrostatic repulsion, but the total energy must remain the same. This means that the kinetic energy has to decrease, and so the alpha particle slows down.

The potential energy continues to increase as the two get closer together, and eventually it's equal to the initial total energy of the alpha particle. At this point, the kinetic energy must be zero, so the alpha particle comes to a complete stop for an instant. It's still feeling the repulsive push from the nucleus, though, so it will almost immediately begin moving away, with the potential energy decreasing and the kinetic energy increasing as it shoots back out the way it came.

In this energy picture, the alpha particle behaves very much like a ball rolling up a hill: as it comes up the hill, it slows down, and eventually stops and turns around. The maximum height it can reach—and thus the minimum separation between the particle and the nucleus—is determined by the initial energy of the particle coming in. When it rolls back down the "hill," it gets back to the starting point with that same energy, moving at the same speed it started with, but in the opposite direction.

In this classical energy picture, an incoming particle with energy less than the height of the peak—at least 8.6 MeV for uranium, according to Geiger's experiments—can't possibly reach the interior of the nucleus where the strong force kicks in. And, by the same logic, a particle starting *inside* the nucleus can't get out unless it has a total energy above the height of the barrier—a particle with any lower energy will hit a wall where the rapidly increasing potential energy equals the total energy it started with, bringing it to a stop and sending it back into the nucleus.

The contradiction between Geiger's scattering experiments and the alpha decay of uranium, then, makes absolutely no sense from the standpoint of classical physics. An alpha particle with just enough energy to escape the strong interaction should start at the top of the "hill" and roll down, emerging at an energy basically equal to the height of the peak. That means that at a minimum, an alpha that barely squeaked out should reach the outside world with 8.6 units of energy, and one that easily escaped should have more. And yet, somehow, uranium naturally decays by emitting alpha particles with less than half that energy.

Even as the development of quantum physics solved many other mysteries of the atom, the problem of alpha-particle energies remained a vexing one. It was finally cracked in 1928 by a young Russian physicist, George Gamow, who realized that thanks to the quantum nature of the alpha particles, they don't *need* to have enough energy to escape the nucleus: they can tunnel their way out.

QUANTUM TUNNELING

It's fitting that the problem of how an alpha particle escapes the nucleus when it has too little energy to do so was solved by George Gamow, as he was later to make an improbable escape of his own. Gamow was born in the Ukraine and began his career in the universities of the Soviet Union. As Joseph Stalin came to power and the regime became more oppressive in the early 1930s, Gamow decided he needed to get out. After two abandoned attempts involving kayaking across open water to a Western country,* he and his wife decided to defect while attending the 1933 Solvay conference in Paris. While Gamow was invited, he ordinarily would have had to go alone; instead he brazenly demanded a passport for his wife, as well—in his telling, from Soviet premier Molotov himself—refusing to go without her. Surprisingly, this tactic succeeded, and with the help of Marie Curie and others at the conference, Gamow successfully defected, and he eventually made his way to the United States.†

All of this was arguably enabled by a 1928 visit Gamow made to Max Born in Göttingen to learn about the latest developments in quantum physics. Born was then engaged in some detailed calculations, a type of problem that did not appeal to Gamow, who'd always loved approximate solutions based on intuitive models. Seeking a research

---

*   In one case from Crimea to Turkey, in the other from Murmansk to Norway; both
       times they were turned back by bad weather.
†   He settled at George Washington University in Washington, DC, not far from my
       grad school house with its toast-hating smoke detector.

question more to his tastes in the Göttingen library, he ran across an article of Rutherford's detailing the problem of alpha-decay energies,* and quickly realized the solution. This accomplishment made his reputation in physics, securing him the invitation to the Solvay conference that provided his ticket out of the USSR.

What Gamow realized was that, in energy terms, the combination of the strong nuclear interaction holding the nucleus together and the electromagnetic force pushing alpha particles away made a "barrier potential," turning low-energy particles away from a small region of space. The wave nature of quantum objects like alpha particles allows them to penetrate this barrier for some short distance, and for a thin barrier this gives them a chance of escaping, even when they do not have enough kinetic energy to escape.

In order to talk about the alpha particle and its interactions in quantum terms, we need to describe it in terms of wavefunctions and probability distributions—and when we do, we immediately run into a problem. The probability distribution we would expect from the classical model described above shows a slow increase in probability as the alpha particle comes in toward the nucleus and slows down,† then crashes down to zero exactly at the point where the potential energy equals the total energy. The probability of finding it at any point closer than that turning point is exactly zero.

While this makes perfect sense in classical physics, the wave nature of quantum objects prevents such a sharp cutoff. As we saw when we talked about uncertainty in Chapter 7, a sharp edge to the wavefunction would require the addition of an enormous number of different wavelengths. Such a huge range of wavelengths is incompatible with the idea of a particle of known energy coming in, though. For real particles, the wavefunction can't stop abruptly but must tail off slowly, extending

---

*   At the time, Rutherford was promoting an elaborate model involving a constellation of orbiting alpha particles in the outer region of the nucleus to explain the low energy of alpha particles that escaped the nucleus.

†   The slower speed means it spends more time in that region of space, corresponding to a higher probability of finding it there.

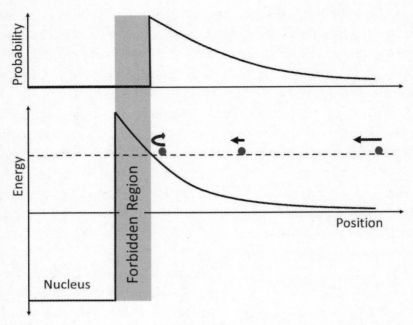

Energy and classical probability for an alpha particle approaching a nucleus from outside. The probability of finding the particle close to the forbidden region increases because the velocity decreases and it spends more time in that area. At the edge of the forbidden region, the probability drops immediately to zero.

some distance into the barrier, and that means there's some probability of finding the particle in places that ought to be forbidden—where the potential energy is greater than the energy it started out with.

Thanks to this tailing off, particles coming in with energies less than the peak of the barrier have some tiny probability of making it inside the nucleus. The probability drops off rapidly* as the particle crosses the forbidden region, but if the energy is not too far below the peak, the forbidden region is thin, and the probability of reaching the inner edge is not zero. And, of course, once it's there, the strong interaction kicks in and holds the alpha particle inside.

---

* The exact shape of the wavefunction depends on details of the potential energy, but it's basically an exponential decay as the particle moves deeper into the forbidden region.

The odds of this happening are astronomically small, and the loss of such a tiny number of particles would have been undetectable in Geiger's experiments. The same process also works in reverse, though, and means that, for some particles starting on the inside, the "box" of the nucleus becomes slightly permeable. There is a small range of positive energies below the peak of the barrier for which particles can potentially be trapped inside the nucleus in a standing-wave–like state. The probability of finding these particles in the forbidden region is not zero, but tails off as the distance increases. And, crucially, that probability will not be zero at the outer edge of the forbidden region.

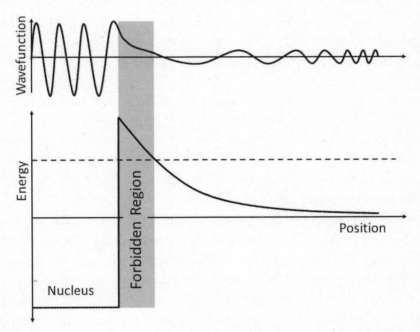

Energy and wavefunction for an alpha particle tunneling out of a nucleus in the Gamow model. The wavefunction decreases exponentially as it moves through the barrier, giving it a small probability of reaching the outside.

Each time the alpha particle encounters the barrier, then, there's some small chance that it escapes. And while an alpha particle being shot at a uranium atom only encounters the barrier once, an alpha particle bouncing around inside the nucleus hits the barrier a lot—physicists

Edward Condon and Ronald Gurney at Princeton, working on the same problem as Gamow, estimated 10^{20} times per second. The probability of any individual encounter resulting in escape is incredibly tiny, but given time, the alpha particle will inevitably end up outside the nucleus at rest. At that point it will be pushed out by the electromagnetic interaction and emerge as the product of a radioactive decay, with a kinetic energy less than the height of the barrier.

This process is referred to as "tunneling," because the particles emerge on the far side of the barrier even though they don't have the energy to go over it, as if they'd dug a tunnel from the west side to the east side of the energy "hill." On learning about the seeming paradox of alpha decay, Gamow quickly realized that tunneling—which he had seen described by his Soviet colleagues Leonid Mandelstam and Mikhail Leontovich in 1928—was the solution to the problem, and he worked out a simple model of radioactive decay as a tunneling process, imagining a radioactive nucleus as a collection of trapped alpha particles with some chance of tunneling to freedom. Gamow's analysis showed that the decay lifetime for a given element should decrease exponentially as the energy of the emitted alpha particles increased, which explained earlier experimental observations by Geiger and John Mitchell Nuttall extremely well.

Gamow's model makes sense of the energy discrepancy revealed by Geiger's experiment, and also explains a number of other properties of alpha decay. The tunneling process is inherently probabilistic—it gives a tiny probability for the particle to escape each time it encounters the barrier, but it can't predict definitively when that escape will happen. This explains one of the signature features of radioactivity, demonstrated by Rutherford in the early 1900s—that the radioactivity of a given sample decays over time at a characteristic rate, its "half-life." The half-life of an element is a statistical quantity, the time after which, on average, only half of the atoms in the initial sample will remain in the initial state; after a second half-life, one-fourth of the initial atoms remain undecayed, and so on. This is exactly what you'd expect for a random decay with some probability, and Gamow's model explains why alpha decay is such a process.

The tunneling model also explains why alpha decay occurs naturally only in very heavy elements. For an alpha particle to be able to tunnel out, it must exist inside the nucleus with its kinetic plus potential energy in the small range of energies that are greater than zero but smaller than the height of the barrier. Thanks to the powerful attraction of the strong interaction, though, most of the allowed states for a particle inside the nucleus are standing-wave–like states with a negative total energy. Such particles have nowhere to tunnel to: there's no region outside the nucleus where they're not forbidden. These permanently trapped alpha particles are responsible for the stable nuclei that make up most of the periodic table.

As in so many other cases, though, Pauli exclusion comes into play, especially for heavy elements. As we add more and more particles to make a heavier nucleus, they fill up the low-energy states. For sufficiently heavy elements, the last few particles added to the nucleus are forced to occupy states with positive total energy, where tunneling can take place. Thus, alpha decay is a phenomenon seen only in heavy elements.

Nearly simultaneously with Gamow's realization in Göttingen, Condon and Gurney in Princeton hit on the same concept for explaining alpha decay. Gamow's approach was a bit more detailed, though: he worked out an excellent approximation of the tunneling rate for a given element and alpha-particle energy that allowed him to make quantitative predictions more easily. As a result, one of the relevant quantities used to determine rates of radioactive decay is known today as the "Gamow factor." The tunneling model was an overnight success, rapidly displacing several more baroque explanations that had been suggested for the alpha-decay energy problem. Gamow's breakthrough quickly established him as an important person in the rapidly developing field of quantum physics, setting up his own eventual escape from Stalin's USSR.

SUNSHINE AND SPLIT ATOMS

The physics of tunneling has cropped up already in our story of an ordinary morning—in the first chapter, when we talked about the sun—though we didn't draw much attention to it. For fusion to take place, two protons must come close enough for the strong nuclear interaction to bind them together, and two protons colliding inside the sun experience the same sort of interaction energy as an alpha particle approaching an atomic nucleus: repulsive at medium range, and attractive at short distances where the strong interaction kicks in. It's a simple matter to estimate the energy required for a proton to pass over the resulting energy barrier—it's just the potential energy due to electrostatic repulsion for two protons separated by the width of a nucleus—which would correspond to a temperature of around fifteen billion kelvin. As hot as the core of the sun is, it's not *that* hot—more like ten million kelvin, a factor of 1,500 too small for direct fusion to occur.

The fusion reactions that power the sun occur through tunneling: even though the protons don't have the energy required to get close enough for the strong interaction to stick them together, their quantum nature gives them some probability that they can tunnel through the barrier and fuse. This is a fantastically unlikely occurrence, but there are so many protons present in the sun that it happens often enough to keep our most important star hot and shining.

When Gamow put forth the idea of alpha decay as a tunneling process, some experimental physicists working with Ernest Rutherford (who was by this time head of the Cavendish Laboratory in Cambridge), notably John Cockcroft and Ernest Walton, quickly realized that the opposite process also ought to be possible. A charged particle shot at the nucleus of an atom will have a small chance of penetrating the barrier and reaching the interior of the nucleus, and under the right circumstances this might be able to knock some particles out. Getting particles inside the nucleus had long been a goal of Rutherford's lab, but the energy needed to get particles over the repulsive barrier was too high to reach in practical experiments using naturally occurring radioactive sources. Gamow's tunneling model, though, suggested that they might not need such a high energy after all, bringing the interior of the nucleus within reach of artificially produced high-energy particles.

Cockcroft and Walton set to work on making a particle accelerator to produce high-energy protons, and in 1932, they managed to successfully penetrate the nucleus of a lithium atom.* This is an extremely rare occurrence—they estimated about one in a billion protons made it in—but adding that extra proton to the nucleus of lithium creates an unstable isotope of beryllium, which rapidly splits into two alpha particles, providing a clear signal of their success. Cockcroft and Walton were the first physicists to split an atom and shared the 1951 Nobel Prize for the achievement. Their accelerator, along with the Van de Graaff accelerator and the cyclotron developed by American physicists Robert Van de Graaff and Ernest Lawrence at about the same time, kicked off a new era of experimental nuclear physics, leading to the ever-larger particle accelerators that would reveal the physics of the Standard Model.

At around the same time in Europe, Irène and Frédéric Joliot-Curie discovered "artificial radioactivity." The Joliot-Curies had narrowly missed out on the discovery of the neutron (they produced evidence of neutrons without realizing what they had done—enabling yet another Rutherford associate, James Chadwick, to complete a series of experiments identifying their new particle), but in studying the behavior of neutrons, they discovered that previously inert elements exposed to neutrons sometimes become radioactive. Within a few years, physicists were able to produce all manner of radioactive elements not found in nature; the Joliot-Curies shared the 1935 Nobel Prize.

The neutron absorption process discovered by the Joliot-Curies isn't the same as the tunneling mechanism identified by Gamow and exploited by Cockcroft and Walton—neutrons aren't charged, so they don't need to tunnel into the nucleus in the same way. It does, however,

---

* This was, of course, a long and involved process, entertainingly chronicled in Brian Cathcart's *The Fly in the Cathedral*, which provides an excellent picture of life in the Cavendish during Rutherford's heyday there.

bring us back to where we started, with the operation of a modern smoke detector. The americium-241 used as an ionization source in a typical smoke detector is artificial, made when plutonium atoms absorb neutrons from a nuclear reactor. The half-life of americium is a bit more than four hundred years, making it ideal for smoke detectors: they'll continue ionizing air molecules for far longer than most of the houses they protect will remain standing.

The radioactive decay of artificially created elements is essential for medical imaging technologies. Radiologists can measure the functioning of various organs by introducing radioactive isotopes with short half-lives and tracking their progress through the body with radiation detectors outside. Radioactive technetium added to food is used to track how rapidly material moves through the digestive system, for example. Specific elements can also be used to test particular organs: the thyroid gland uses a lot of iodine, so radioactive iodine isotopes introduced into the body will tend to concentrate there, allowing doctors to check that the thyroid is functioning properly and take images of the glands using gamma-ray detectors.

Artificial radioactivity helps not only to detect diseases, but also to treat them. Medical physicists treat cancers by implanting "seeds" containing artificial elements emitting beta or alpha particles to kill cancerous cells. Depending on the type and location of the tumor, physicists can choose from a wide variety of isotopes to find ones whose half-life and decay energies will do maximum damage to tumors while minimizing harm to normal tissue.

Quantum tunneling also finds numerous applications in laboratory contexts that are far removed from the everyday. One of the most impressive uses is the scanning tunneling microscope, invented in 1981 by Gerd Binnig and Heinrich Rohrer, which uses the tiny current due to tunneling between a sharp metal tip and a surface to measure distances between tip and surface that are smaller than the height of a single atom. This allows physicists to map out the structure of materials atom by atom, and even build atomic-scale structures by pushing atoms around on surfaces to make interesting patterns.

It may come as a surprise, then, to learn that these phenomena are also used in something as ordinary as a smoke detector. If it is a surprise, hopefully it's a pleasant one—exotic physics being put to use to protect lives and property. The next time a smoke detector warns you of a burned meal before it becomes a serious threat—or just that you've toasted bread in an unacceptable manner—some of the credit (or blame) belongs to sneaky alpha particles tunneling their way out of unstable nuclei.

CHAPTER 11

ENCRYPTION: A FINAL BRILLIANT MISTAKE

*My email is mostly from students requesting homework help, plus a couple of receipts and tracking notices from **online purchases** . . .*

While the concept of internet commerce seemed hopelessly exotic barely twenty years ago, buying things online has now become so much the norm that venerable chain stores have been pushed to and even over the brink of bankruptcy by the growth of web-based retail. You can buy almost anything on the internet these days, and for some people even a quick run out to buy milk has been replaced by pointing a web browser at an online grocery service.

Of course, e-commerce would be all but unthinkable without the ability to encrypt messages, enabling a customer to send credit card information to a retailer without worrying that it's being shared with the entire world. Vast sums of money have been spent on technologies to secure commercial transactions over the internet, and the development of successful methods for sharing financial information is largely responsible for the explosive growth of online markets.

This may seem an odd topic for a book focused on the quantum, because at the moment, the security of online transactions is guaranteed by purely classical means. But as we complete our exploration of the physics of everyday reality, we'll take one brief detour into the speculative. The quantum cryptography technology I'll describe in this chapter is not widely used . . . yet.

These techniques are very real, though, and becoming more practical every day. In the fall of 2017, researchers in Beijing and Vienna demonstrated quantum-secured communication via a Chinese satellite, opening a research conference with a phone call between China and Austria that was encrypted with a quantum key. Widespread global quantum communication is not far off, despite the fact that it has its roots in some of the most exotic physics ever discovered.

Quantum cryptography draws on the idea of "entanglement," arguably the most troubling of all the weird properties of quantum mechanics. Quantum entanglement establishes connections between particles at great distances, an idea that Einstein famously derided as "spooky action-at-a-distance." Numerous experiments since the 1970s have demonstrated the reality of this phenomenon, though, forcing physicists to grapple with the deeper meanings of space, time, and the transfer of information.

At first glance, the questions entanglement raises may seem primarily philosophical, but in fact they have deeply practical applications. If you're trying to transmit messages from one person to another without anyone else being able to read them, this "spooky" connection between entangled particles turns out to be exactly what you want.

THE SECRET TO KEEPING SECRETS

The central problem of cryptography has probably been around as long as written language. The most obvious way to keep secrets is, of course, to only share them face to face, but in-person meetings are not always practical. One solution is the use of codes: writing a message in a way that's intelligible to the intended recipient, but gibberish to anyone who intercepts it.

There are numerous ingenious code systems dating back thousands of years, but we're interested in modern security, which is best understood in terms of math. In modern cryptosystems, the secret message is converted into a string of numbers, and then some mathematical operation is performed on those numbers by the sender, resulting in a different string of numbers that is sent openly to the recipient. If the recipient knows exactly what was done, they can undo it, recovering the original message; anybody else will be left with a string of nonsense.

To give a concrete but rudimentary example, imagine doing the conversion from letters to numbers by simple substitution: A = 01, B = 02, all the way to Z = 26. If we want to encode the word "BREAKFAST," we end up with

B	R	E	A	K	F	A	S	T
02	18	05	01	11	06	01	19	20

For a mathematical operation to obscure this, we take a random string of 1s and 0s as our cipher key, one for each letter of the message. We then combine the two, adding one to the original number if our key has a 1 in that spot, and subtracting one if our key has a 0.

B	R	E	A	K	F	A	S	T
02	18	05	01	11	06	01	19	20
0	1	0	0	0	0	1	1	0
01	19	04	26	10	05	02	20	19
A	S	D	Z	J	E	B	T	S

A person receiving the cipher text "ASDZJEBTS" without knowing the code would most likely conclude that the sender's cat was walking on the keyboard again. If the intended recipient has the key and knows the appropriate sequence of operations, though, they can reverse the encryption—adding one for a 0 in the key, subtracting one for a 1 in the key—and recover the original message.

This simple cipher illustrates the basic principle, which is also the chief problem: it depends on both sender and recipient knowing the right sequence of operations to apply, according to a shared key—in this case, 010000110. If the recipient doesn't have the same key as the sender, they're no more able to decode the message than some random eavesdropper.

The simplest way around this is to use a single key all the time, so both sender and receiver only need to share and remember one special set of digits. Unfortunately, with enough ciphered message text to work from, mathematical analysis can determine the key and recover the secret message, given enough time. "Enough time" may be a lot—for a long enough key, the time required to be sure of deciphering a message with current methods on existing computers can be longer than the age of the universe. This is what most internet messages rely on: they use a single shared key with enough digits that it's exceedingly unlikely that anyone will figure it out fast enough to do harm. This sort of cryptography is vulnerable to improvements in computing power or new mathematical techniques, though—a person with a better decryption program and bad intentions can potentially decrypt vast amounts of material.

A more secure method is to have a list of random numbers to use as keys—a so-called "one-time pad"—and use a new one to encode each message, but this creates additional logistical hassle for the sender and receiver. Each must have access to some large shared list of random numbers, and the longer the list of numbers that needs to be shared, the harder it is to keep them secret.* It can also be difficult to securely

---

* It's easy to memorize or hide a short list of numbers, but the longer the list, the harder it is to keep track of in a non-obvious way. The problem is rather like the way most people can easily remember short but not very secure passwords, but longer, more secure strings of numbers and letters end up written on Post-it notes stuck to the monitor, completely defeating their purpose.

replenish the list after many messages if the sender and recipient are in places where they can't easily meet.

The ideal system for this sort of cryptography would be one that somehow generated random numbers on demand. But while there are plenty of random processes either sender or recipient could use to generate a useful key, if they're doing the generation in two different places, the numbers produced will necessarily be different, and thus useless for encoding text. The need for the sender and receiver's numbers to be identical makes on-demand key generation all but impossible.

At least, all but impossible *in classical physics*. Quantum mechanics, though, provides a loophole that allows you to generate a truly random number that is nonetheless shared by two people in two different locations. It works thanks to one of the thorniest philosophical issues raised in quantum physics, the one that drove Einstein out of the field he had helped invent.

DICING WITH THE UNIVERSE

One of the most frequently shared Einstein quotes is usually rendered as something like "God does not play dice with the universe." This traces back to a remark first made to Max Born in a 1926 letter: "[Quantum mechanics] says a lot, but does not really bring us any closer to the secret of the 'old one.' I, at any rate, am convinced that He does not throw dice."*

The fundamental issue here has to do with the probabilistic nature of quantum mechanics, first stated by Born: quantum wavefunctions tell us only the *probability* of particular measurement outcomes. If we repeat an experiment many, many times, and aggregate all the results, the wavefunction will be an excellent description of the full range of results. Knowing the wavefunction, however, does not allow us to

---

* The original was, of course, in German: "Die Theorie liefert viel, aber dem Geheimnis des Alten bringt sie uns kaum näher. Jedenfalls bin ich überzeugt, dass der Alte nicht würfelt."

predict the exact outcome of any particular run of the experiment; as far as we can tell, the result of a single experiment on a quantum particle is completely random.

This randomness poses a serious philosophical problem. Probability per se is not the issue, even for Einstein himself—as we've seen, some of his greatest contributions to physics involved using statistical methods to predict the behavior of vast numbers of particles without needing to consider the details of any individual particle's behavior. In those cases, though, he could presume that the randomness was just covering for a lack of *knowledge* about the detailed interactions. A deeper theory that would predict specific results for individual particles remained a possibility, in which case the statistical methods would just be a convenience, a tool for avoiding the impossible task of calculating the details of the interactions between huge quantities of individual particles. We do this with purely classical systems all the time—knowing the initial position and velocity of a roulette ball and wheel would in principle allow you to predict exactly where the ball will stop, but in practice, that calculation is too difficult, and instead we can treat the game as purely random and discuss the outcome in terms of probability.

As quantum mechanics began to emerge, though, it became clear that, in quantum mechanics, randomness is *fundamental*. The inability to predict the outcome of a single quantum experiment isn't just some technical glitch, it's inherent. In the quantum theory formulated by Heisenberg and Schrödinger, and interpreted by Bohr and Born and Pauli, it simply does not make sense to talk about specific properties of individual particles. The Heisenberg uncertainty principle (Chapter 7) isn't describing a technical issue with the way we measure position and momentum; it reflects the fact that it's simply impossible to have a well-defined position and momentum for a particle that also has wave nature.[*]

---

[*]   The de Broglie–Bohm pilot wave approach is an alternative approach to quantum theory where individual particles *do* have definite properties, but are guided by an additional field that takes on most of the weirder properties associated with

The younger generation of physicists, Pauli and Heisenberg and their cohort, were largely willing to accept this as the cost of doing business, and reveled in the ability of the new theory to accurately predict the results of experiments that had baffled physicists for years. Some older physicists, though, were deeply troubled by this fundamental randomness, and sought a replacement theory that would be more deterministic.* This group included some of the physicists who'd been instrumental in the launching of quantum mechanics in the first place, most notably Einstein and Schrödinger.

This is the context of Schrödinger's infamous cat thought experiment: he was highlighting what he saw as a problem for quantum theory, relating to this fundamental indeterminacy. While the question he raised did not deter the further development of quantum mechanics, the arguments it inspired have helped generate new and productive areas of research. Einstein's objection, in the form of another thought experiment, was to prove even more fruitful.

QUANTUM PHYSICS AND BETTERIDGE'S LAW OF HEADLINES

In the late 1920s, Einstein had a celebrated series of debates with Niels Bohr about interpretations of quantum physics, at the Solvay Conferences of 1927 and 1930. The initial arguments focused on the uncertainty principle, which Einstein initially objected to because it went against classical intuition. While Einstein eventually reconciled himself to the idea of the uncertainty principle, and moved on to a

---

quantum particles. The specific initial properties of any individual particle are still randomly determined and impossible to measure, though, so the results of a single run of a quantum experiment remain unpredictable.

* A deterministic alternative to quantum physics remains a topic of interest for a handful of researchers, most notably Gerard 't Hooft (who shared the 1999 Nobel Prize in Physics for work on the Standard Model), but subsequent generations mostly have followed the lead of Pauli and Heisenberg and prefer, in the tongue-in-cheek phrasing of David Mermin, to "shut up and calculate."

deeper objection, Bohr continued to interpret his arguments in that light, meaning that a lot of their celebrated arguments are actually two brilliant physicists talking past each other.

Einstein's final and most significant contribution to the still-ongoing argument about the foundations of quantum mechanics came in a 1935 paper written with his younger colleagues Boris Podolsky and Nathan Rosen. The "EPR" paper caught Bohr and many other physicists who were used to regarding Einstein's arguments as uncertainty-based totally by surprise, because it more clearly explained his real objection, and pointed at a deeper issue with quantum theory.*

The paper is titled "Can Quantum-Mechanical Description of Physical Reality Be Considered Complete?" There's an old joke among journalists, "Betteridge's Law of Headlines," holding that any story whose headline is a question can be answered with the single word: "No." The EPR paper is no exception: Einstein and his colleagues considered an unusual physical system to argue that quantum theory as developed and interpreted by Bohr and his colleagues in Copenhagen could not capture all of physical reality. This was the formal introduction of "entanglement" to physics,† and the concept has troubled physicists ever since.

The original EPR argument involves the position and momentum of two particles, but the argument is clearer when applied to a two-state system like the spin of an electron. As we saw in the Stern-Gerlach experiment (Chapter 6), you can use a magnetic field to separate a bunch of electrons into two groups: one with the spin pointing "up" (along the magnetic field), the other with the spin pointing "down" (opposite the field).

The orientation of the Stern-Gerlach magnet is an arbitrary thing, though—up and down are not well-defined directions in space, and

---

*   It's not a perfect presentation, though. In his later years, Einstein reportedly said that he was dissatisfied with the final wording of the EPR paper, which was largely written (in English) by Rosen.
†   The term "entanglement" was coined by Schrödinger, who shared Einstein's misgivings.

you could perfectly well tip the whole apparatus on its side and get the same basic result: half of the electrons will be "spin-left," and half "spin-right." If you start with a random sample of electrons, and a randomly chosen magnetic field direction, you'll always get two groups. If you select one of the groups and repeat the measurement, passing it through the same magnetic field a second time, the results remain the same as well: all the spin-up electrons will stay spin-up (or the spin-left ones will stay spin-left), and vice versa.

An obvious extension of this experiment is to take one of the two groups separated by one Stern-Gerlach magnet—spin-up, say—and feed it into a second magnet with a *different* orientation—say, left-right. When you do this, you'll again find two groups—for instance, half of the spin-up electrons will be spin-left, and half spin-right. The same is true if you do left-right first, then up-down, or any combination of two magnet sets where the second is rotated by 90 degrees.

So far, so good, but things get weird when you add a *third* magnet. Common sense would seem to say that if you take the group of electrons that were spin-up in the first magnet and spin-left in the second, then pass them through a second up-down magnet, you should find all of them in the spin-up group. After all, they were already measured to be spin-up.

That's not what happens, though. The electrons that were first spin-up and then spin-left will separate into two equal groups: half spin-up, and half spin-down. Somehow, the process of measuring these electrons as spin-left has erased the original spin-up result, returning you to a random outcome for the up-down measurement.[*]

In the mathematical description of spin worked out by Pauli, the reason for this is simple: the up-down and left-right measurements of an electron's spin are complementary to each other in the same way

---

[*] Information about the original state is completely lost only for the case where the magnets are rotated by 90 degrees. If you choose an intermediate angle, you get two groups, with different probabilities—rotating the second magnets by 60 degrees from spin-up, say, gives a 75 percent–25 percent split between spin-right and spin-left. This will become important later.

that measurements of its position and momentum are. They're subject to an uncertainty principle–like relationship, and it simply does not make sense to talk about the up-down and left-right states of an electron's spin as having well-defined values at the same time.

The EPR paper, though, uses a system of two particles to argue that this quantum indeterminacy cannot be a complete description of reality. They imagined a state of two particles whose individual state was indeterminate, but whose combined state had a definite value. In the spin framework, this would mean knowing that the two particles have opposite spins—one up and one down, or one left and one right—but not which is which. (This is not difficult to arrange—you get this kind of state, for example, from a reaction that breaks a diatomic molecule in two.) They then imagined separating these two particles by a substantial distance before measuring their individual properties.

The correlation means that when the scientist in possession of particle A (traditionally named "Alice" in discussions of cryptography) measures spin-up, they can predict with absolute certainty that particle B (held by Alice's colleague Bob) will be found to be spin-down. They can't say in advance which will be which, but the correlation between the results is absolute, and knowing the state of one particle instantly tells you the state of the other.

If we stick with a single measurement, this isn't particularly surprising even from a classical standpoint. If I take the queen of spades and the jack of diamonds from a deck of cards, and mail them in sealed envelopes to different locations, when Alice opens her envelope to find the queen of spades, she knows instantly that Bob has the jack of diamonds, no matter where he is. The randomness in this case reflects merely a lack of knowledge about the state, not any inherent indeterminacy: each envelope contains a specific card the whole time it's making its way through the postal system—we just don't know which is which.

In the spin case, though, we're not restricted to a single measurement, but could choose between two complementary measurements—if Alice had instead chosen to measure left-right, an outcome of spin-left would let them know with absolute certainty that Bob had spin-right. The randomness here is not a simple lack of knowledge in the classical sense, but rather a more fundamental indeterminacy. It's as if I mailed two cards from a deck, and opening the envelope from the top would reveal either the queen of spades or jack of diamonds, while opening the envelope from the end would reveal either the ace of hearts or the two of clubs. In this case, we're not only unsure of what specific card is in each envelope, it's not even clear what the options are until the envelope is opened.

But, as Einstein, Podolsky, and Rosen pointed out, the particles have no way of knowing beforehand which measurement to expect, left-right or up-down, and there's no restriction on the timing of the measurements that would allow for a message to pass from A to B to tell the other particle which outcome to choose. And yet, the correlations between measurements must be maintained. To Einstein, this suggested that all of the possible measurement outcomes *must* be determined in advance, each particle carrying with it a set of instructions for what result to show for any given measurement. Such a list of results, though, would go against the idea of quantum indeterminacy—each individual particle really *would* have a definite state the whole time, with the measurement results determined by some hidden variable not described in quantum mechanics, but potentially able to be determined with some deeper, more complete theory.

The only alternative would be what Einstein derisively referred to as *"spukhafte fernwirkung,"* a "spooky action-at-a-distance" communicating the result of Alice's measurement to Bob's particle at speeds far exceeding the speed of light. Such a linkage between widely separated particles would violate basic intuitions about space, time, and information as described by the theory of relativity. That sort of "non-local" interaction would create such enormous problems for classical physics—if you could send information faster than light, you could even create a paradox where effects happen before their causes—that Einstein rejected it out of hand.

EINSTEIN TO BELL TO ASPECT

The EPR paper, in the words of one of Bohr's close colleagues, Leon Rosenfeld, "came down upon us as a bolt from the blue." The Copenhagen circle of physicists had not anticipated this line of argument, and struggled to understand it. Bohr rushed out a paper in response with the same title, "Can Quantum-Mechanical Description of Physical Reality Be Considered Complete?", but this mostly served to muddy the waters; Bohr was not a clear writer in the best of times, and was caught badly off guard by the EPR thought experiment.

Over time, the response coalesced into a challenge to one of the central premises of the EPR argument, namely that the measurement at point A is made "without in any way disturbing" the measurement to be made at point B. In Bohr's words, the fact that the two particles are entangled into a single quantum state means that Alice's measurement exerts "an influence on the very conditions which define the possible types of predictions regarding the future behavior" of Bob's particle. According to the Copenhagen interpretation, the complete quantum description of reality inherently incorporates all the measurements that will or might be made at widely separated locations.

This approach to entanglement didn't really make anybody happy, but the situation seemed so arcane and artificial that most physicists didn't give it much thought. Quantum mechanics was spectacularly successful at calculating the properties of a huge number of interesting systems, and most physicists focused their energy on those calculations, not an odd philosophical dispute between Einstein and Bohr that nobody could address in an experiment. Both camps were in agreement about what the measurable results of an EPR-type experiment would be; they disagreed only about the "why" of those results—whether the outcome was truly indeterminate but entangled, or determined in advance by hidden variables. Bohr's view drew extra support from an assertion by John von Neumann that a "hidden variable" theory was mathematically impossible; von Neumann turned out to be flatly wrong on this point, but he was so respected that many physicists who were inclined to Bohr's view simply accepted his claim without checking the math.

This muddled philosophical impasse remained for almost thirty years without a breakthrough. Einstein and Schrödinger basically gave up on quantum theory, moving on to other fields,[*] and quantum mechanics continued to develop on the lines laid out by Bohr and his Copenhagen colleagues. In the mid-1960s, though, an Irish physicist named John Bell took a careful look at the Einstein, Podolsky, and Rosen argument, and realized that there was a way to experimentally distinguish between the "local hidden variable" theories they preferred and the orthodox quantum explanation.

The key to Bell's trick is to look at what happens when Alice and Bob make *different* measurements. If the two spin detectors are set to make the same measurement—both looking for up-down, or both left-right—then the results will be simply correlated, and there's nothing more you can do with that. If they're looking at different properties, though—one up-down and the other left-right, say—then there's some probability of getting each of the possible combinations. And the range of possible probabilities is different for the local hidden variable theories than it is for quantum mechanics.

The essence of the local hidden variable approach is that each particle must carry with it a set of instructions as to what result it should return for each of the possible measurements that might be made on it. To make this more concrete, we can assign values of "0" and "1" to the two different possible outcomes ("1" for spin-up and "0" for spin-down, say), and consider three different possible settings for the angle of the detector relative to the up-down direction. (Three measurement options is the smallest number that provides enough mathematical complexity to illustrate Bell's theorem; in reality, there are an infinite number of possible choices, requiring tricks from calculus to handle the enumeration.) A local hidden variable theory then allows

---

[*] Einstein devoted the last decades of his life to a fruitless search for a theory that would combine gravity and electromagnetism into a unified field. Schrödinger worked on field theory as well, and also wrote an influential book on the physics of living systems.

pairs of particles to exist in eight possible states, which we can enumerate in a table:[*]

	A 1	A 2	A 3	B 1	B 2	B 3
I	1	1	1	O	O	O
II	1	1	O	O	O	1
III	1	O	1	O	1	O
IV	1	O	O	O	1	1
V	O	1	1	1	O	O
VI	O	1	O	1	O	1
VII	O	O	1	1	1	O
VIII	O	O	O	1	1	1

Each row shows a possible state for a particle pair, and the measurements returned for each detector setting. The "A" columns indicate the results of the measurements made by Alice at each of the three settings, the "B" columns those made by Bob. Any pair of entangled particles used in the experiment must be in one of these eight states, chosen at random.

To understand Bell's argument, we put ourselves in the role of "Setter of Variables," choosing the state of each entangled pair in an attempt to match the predictions of quantum mechanics. We're free to adjust the probability of each of these eight states occurring, subject to the constraint that a collection of repeated measurements by any single detector at any of the settings must always have a 50 percent chance of returning 0 and a 50 percent chance of returning 1.

As you can see, when both detectors have the same setting, the results are always opposite, reflecting the entanglement between the particles, so that part of the variable-setter's job is easy. As Bell pointed out, though, a trickier question to consider is what happens when the two detectors are rotated to different angles. We want our hidden-variable approach to match the quantum predictions, whatever

---

[*]   This approach to illustrating Bell's theorem ultimately traces back to David Mermin.

they may be, so we need to work out the maximum and minimum probability of getting opposite results at A and B for any pair of different settings.

It's relatively easy to see how to make the maximum probability outcome, which is 100 percent: simply put half of the entangled pairs in state I and the other half in state VIII. For each of those states, no matter how Alice sets her detector, a 1 for her will be paired with a 0 for Bob, and vice versa.

To get the minimum probability, we obviously need to exclude those two states; if you look closely at the remaining six, you will see that there are always exactly two states that give opposite results for any particular pair of detector settings. If we use the combination A1 and B2, states II and VII give opposite outcomes, for example; if instead we picked A2 and B3, states IV and V would do the job. If those six states are equally likely, as they must be to ensure a 50/50 chance of 0 or 1 for each individual detector, we have a one-in-three chance of getting opposite results.

The probability of getting opposite results with different settings, then, must range from a maximum of 100 percent to a minimum of 33 percent. As Setter of Variables, we can make our local-hidden-variable source match the behavior of quantum-entangled particles for any scenario, provided that the probability of opposite measurement results never drops below one in three.

So, what is the quantum prediction that the Setter of Variables needs to match? In the quantum picture, the measurements are not independent: in one way of speaking about it, when Alice sets her detector to A1 and gets a result of 1, Bob's particle is definitely placed into the 0 state *for that detector setting*. If the entangled particles are spins, the exact probability of Bob getting a 0 for his particle at a different detector setting will then depend on the exact angle between the settings. If we know that Bob's particle is in a state that will give a result of 0 for the angle that corresponds to Alice's setting A1, the probability of Bob detecting a 0 at setting B2 will be 100 percent if B2 is the same as A1, and decreases as B2 is rotated to a larger angle away from A1. Working through the details shows that the probability can be as low as 25 percent (for an angle of 60 degrees between detectors).

So, the Setter of Variables has an impossible task: for some combinations of detector settings, the probability of opposite measurements predicted by quantum physics is lower than the minimum probability that can be arranged using local hidden variables. What's more, a careful experiment can readily distinguish between a probability of 25 percent and one of 33 percent, allowing physics to settle the argument between Bohr and Einstein once and for all.

Of course, reality is more complicated than our eight-state toy model, but then, so was Bell's argument. Bell considered a much more general case, and proved an airtight mathematical theorem showing that for any EPR-type experiment, there will always be some choice of detector settings that makes predictions that local hidden variable theories simply cannot match.

Bell's initial papers about the EPR experiment didn't attract wide notice, but caught the interest of some physicists who decided to do the experiment. An initial test in the mid-1970s led by John Clauser found a result that agreed with the quantum prediction, though with weak statistical power. In 1981 and 1982, a young French physicist named Alain Aspect did a series of experiments that are widely regarded as definitive, getting results that agreed with the quantum limit, and closing some obvious loopholes that might've allowed a local hidden variable theory to mimic the quantum result.[*] Over the last thirty-five years, numerous additional "Bell test" experiments have been carried out, and all of them show the same thing: the quantum prediction is correct. The local hidden variable approach Einstein, Podolsky, and Rosen favored cannot be the correct description of our quantum universe.

---

[*]   The details of Aspect's experiments are fascinating, but too long to go into here. For more detail on the experiments, see *How to Teach Quantum Physics to Your Dog*. The story behind the Clauser and Aspect experiments is also fascinating, and well told in David Kaiser's *How the Hippies Saved Physics*.

QUANTUM CRYPTOGRAPHY

To physicists, the most fascinating thing about the EPR argument and Bell's theorem is what it tells us about the fundamental nature of the universe—these "spooky" correlations between entangled particles are very real, and confirmed in countless experiments. This means that distant points can have a quantum connection between them, which seems to run counter to our intuition that widely separated locations are, in fact, separate. Working out the details of this fundamental non-locality and what prevents it from manifesting more widely and upending our normal reality is a fascinating subject occupying a small but active community of physicists and philosophers.[*]

In this book, though, we're mostly concerned with how aspects of quantum physics impact ordinary, everyday activities, and as fascinating as quantum foundations research may be, perhaps the most notable thing about quantum entanglement is its *absence* from everyday reality. In an everyday context, we simply don't see it producing obvious practical effects.

There is one extremely practical application of quantum entanglement, though: its use in quantum cryptography. You can see this by looking at the raw data for any experiment on entangled particles: each individual measurement at point A will give a 0 or a 1 at random, but the scientist making those measurements will know with absolute certainty that their compatriot at point B making the same measurement has the opposite. The process allows two widely separated people to generate two lists of perfectly random numbers that are nevertheless perfectly correlated. That's exactly what you need to encrypt and decrypt secret messages.

A twist based on real Bell-test experiments also allows our secretive physicists to rule out the possibility of eavesdropping, by switching between different detector settings while measuring their shared particles. Alice and Bob share a large number of entangled particle pairs

---

[*]   George Musser's book *Spooky Action at a Distance* is a good overview of the history of non-local interactions in physics, and exciting current research in the field.

(which we'll continue to talk about as if they're electron spins), and as they work through the list, they make a random decision whether to measure up-down or left-right. After making all their measurements, Alice openly shares the list of what measurement she made to each spin—not the value, just whether it was up-down or left-right. Roughly half of the time, Bob will have made the same measurements, and their results will be perfectly correlated—a 1 for Alice is a 0 for Bob, and vice versa. If Bob tells Alice which measurements were the same—not the outcome, just which pairs had the same detector settings—they get a set of perfectly correlated random numbers. When Alice finds a 1 in that half of the data, she can infer that Bob has a 0 and vice versa, and they can use those digits to make the key they need to encrypt their message.

The random switching between measurements slows the rate at which they generate bits for their key, but it foils would-be eavesdroppers. To have any chance of stealing the key, Alice's archenemy Eve needs to intercept one of the entangled particles and make her own measurement of its state before sending Bob a replacement particle in the definite state corresponding to her measurement result—if she measures spin-up and gets a result of 1, she prepares a new particle in the 1 state, and sends it on to Bob. Since Eve has no way of knowing what measurement will be made, though, she has to choose her detector settings randomly as well, and this will inevitably introduce errors: if Eve measured up-down while Alice and Bob measured left-right, there's a 50 percent chance that they'll end up with two 1s rather than the 1-0 pair they expect.

Eve's attempt to intercept the key will thus introduce errors, meaning that the attempt to decrypt the message will produce some gibberish characters. More importantly, it allows Alice and Bob to detect Eve's presence—they can measure many more pairs than they need for the key, and then pick some random sections of that list as a test, sharing not just what measurement was made, but the outcome of the measurement. If Alice and Bob find too many cases where the correlation is imperfect, they know Eve is trying to intercept their key, and can take steps to eliminate the threat.

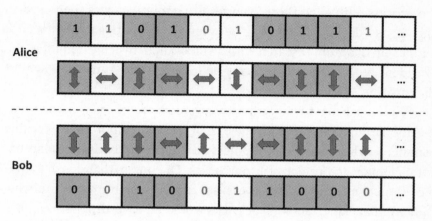

Illustration of quantum key generation. Alice and Bob share entangled spins, and each randomly decides whether to measure spin-up/spin-down or spin-left/spin-right. When their measurement choices match (shaded boxes), a 0 for Alice means a 1 for Bob, and vice versa. If they share what measurement they made for each spin, and keep the results for spins where they made the same measurement, they get correlated random numbers that they can use as a cryptographic key.

In practice, of course, there are many technical details that complicate the basic process just outlined. Real-world quantum cryptography systems use polarized photons as their entangled quantum particles, and reliably sending and detecting single photons can be very challenging. This has been an active area of research since the first proposals in 1984, though, and steady progress has been made. Quantum key distribution using polarized photons sent via optical fiber has been demonstrated at distances of several hundred kilometers, and is reliable enough that commercial systems are available.

The Chinese team mentioned earlier in the chapter has also demonstrated quantum key distribution between ground-based labs and a satellite in orbit. In the fall of 2017, they conducted that first "quantum-secured" international call between China and Austria via a Chinese satellite (named "Micius," after the Latinized name of a Chinese scholar from the fifth century BCE). As Micius passed above the lab in Beijing, they aimed laser pulses at the satellite to generate a key. A short time later, as the satellite passed over Vienna, they repeated the process with a lab there. The resulting joint key was then used to encode and decode a video link between the two cities to open a conference on quantum research with a video call between Chunli Bai, the president of the Chinese Academy of Sciences, and Anton Zeilinger of the Austrian Academy of Sciences.

While quantum key distribution systems are not yet in wide use, it's not hard to believe, given the ever-increasing importance of online commerce, that banks and retailers will someday be using quantum entanglement to protect your purchases. Of course, that doesn't completely guarantee security—there are also research groups studying "quantum hacking," looking at tricks would-be eavesdroppers can use to disguise themselves and steal quantum keys. Quantum mechanics won't end the arms race between those trying to keep secrets and those trying to steal them; it will just shift the fight to new and spookier ground.

A BRILLIANT MISTAKE

Einstein's turn away from quantum physics after his pivotal role in inventing it was long regarded as an unfortunate footnote to a brilliant career. Abraham Pais's magisterial scientific biography of Einstein, *Subtle Is the Lord . . .*, barely touches on the EPR paper, treating it as a brief and unfortunate late-career episode.

Ironically, Pais's book was published in 1982—also the year when Alain Aspect's third experiment using entangled photons was published, widely regarded as one of the best real-world realizations of the EPR scenario. That experiment showed fairly conclusively, thanks to the work of John Bell, that quantum-entangled particles are correlated in ways that simply cannot be explained with the sort of local hidden variable theory that would've satisfied Einstein. Since that time, the stature of the EPR paper has grown enormously. A 2005 analysis showed that the EPR paper was cited just 36 times before 1980, but 456 times between 1980 and 2005. In late 2017, the online article showed more than 5,900 citations.

In the end, the argument presented by Einstein, Podolsky, and Rosen turned out to be wrong, but not boringly so. In fact, it's a brilliant mistake, bringing to light a strange and troubling aspect of quantum physics that had not previously been considered. It's wrong in deep and subtle ways, and working out exactly how and why such a seemingly common-sense approach to physics fails has inspired an enormous amount of progress, in both the philosophy of physics and the technology used to probe the fundamental weirdness of entanglement.

In that sense, then, the EPR paper is not an unfortunate footnote to Einstein's career in quantum physics, but a fitting end to it. He helped launch the field in 1905 with the bold claim that light could be a particle, and the dramatic introduction of entanglement thirty years later was an equally bold stroke, albeit in the opposite direction. Each of those papers, in its own way, transformed our understanding of the universe, showing the deep strangeness that exists in the foundations of our ordinary, everyday reality.

CONCLUSION

We began this book with the observation that most people associate physics with extreme and exotic phenomena: the strange particles brought into fleeting existence in giant particle accelerators, the sudden creation of matter and space-time itself in the Big Bang, the mysterious fate of giant stars that collapse to form black holes. These take place on scales that fire the imagination, with results that defy our everyday intuition of how the world ought to operate.

As we've seen through the course of this book, though, the same physics principles that come into play in those extraordinary scenarios also affect extremely mundane activities like getting out of bed and making breakfast before going off to work. Even as basic a fact of our existence as the stability of solid objects turns out to require quantum theory for an explanation: if not for electron spin and the Pauli exclusion principle, any attempt to make a macroscopic object would end in a catastrophic implosion. Everything that we do, no matter how boringly ordinary, is ultimately rooted in quantum physics.

I hope this book has also made clear, though, that this connection goes both ways: that is, exotic quantum physics is ultimately rooted in very ordinary phenomena affecting the behavior of everyday objects. The entire field began with the deceptively simple question, "Why does a hot object glow *that* particular color?" The changing light from a hot object is so common—whether it's an electric toaster oven, an incandescent light bulb, or the sun itself—that we almost forget it's a phenomenon that needs explaining at all. Thanks to the curiosity of the nineteenth-century spectroscopists who decided to study the color carefully, and Max Planck's brave and bold trick, we were set on the path to the strangest and most powerful theory in physics.

Physicists didn't jump to strange and counterintuitive theories in a single step, however; rather, we were inexorably led there by a chain of reasoning each step of which begins with a phenomenon that's readily observable in relatively unremarkable circumstances. Planck introduced the quantum hypothesis to explain black-body radiation, then Albert Einstein used that idea to explain the photoelectric effect, which led to photon statistics, and then to lasers. Marie Curie dug deeply into radioactivity, which Ernest Rutherford used to discover the nucleus of the atom, which led Niels Bohr to introduce discrete atomic states, which led to ultra-precise atomic timekeeping. Dmitri Mendeleev introduced the periodic table, which led to the idea of electron shells, which led Wolfgang Pauli to introduce the exclusion principle, which turns out to be essential for just about everything.

The story of quantum physics isn't a story of people dreaming up bizarre ideas that only apply in unlikely situations; it's a story of basic curiosity followed through with determination and rigorous logic. And no small amount of courage—the key steps in the process involve bold and startling suggestions from Planck, Einstein, Bohr, Louis de Broglie, and others along the way—ideas that easily could've been (and sometimes were) dismissed as flatly crazy, but that stood up to incredibly exacting experimental tests.

So, the connection between quantum physics and everyday activities is a mutual one. A mundane weekday breakfast would not be possible without quantum physics, and quantum physics would not exist without scientists who looked at the glow of a hot object or the attraction between two magnets and said, "I wonder why *that* happens?"

I hope that, in the end, you'll take a lesson from both sides of this relationship. I hope the discussion of the physics underlying ordinary reality inspires you to look a little more closely at everyday activities, and appreciate their roots in astounding and exotic physics. And I hope the stories of the development of quantum theory will inspire you to follow your curiosity: to ask questions about the world around you, take those questions seriously, and follow them wherever they lead. Most of the time, it turns out to be somewhere amazing.

ACKNOWLEDGMENTS

While a published book usually only lists one name on the cover, there is an enormous number of people whose collective efforts are required to make that final product, who need to be thanked within. In the case of this book, the list starts with my agent, Erin Hosier, for finding this book a place.

Several of the ideas in this book got a "trial run" of sorts on my blog for *Forbes*; thanks to Alex Knapp for the opportunity to write for them. Some of the finer points were also clarified thanks to discussions with other scientists, in particular Ash Jogalekar, Nelia Mann, Doug Natelson, Michael Nielsen, Dave Phillips, Tom Swanson, and Mark Walker. The book is more accurate than it would've been without them; any remaining errors are my fault, not theirs.

Thanks to my editors, Alexa Stevenson at BenBella and Sam Carter at Oneworld, for helping sharpen and focus the argument, and to copyeditors Scott Calamar and James Fraleigh for making it look like I understand grammar and punctuation. And thanks to Jessika Rieck and the rest of the production team for making the finished product look great.

This is my fourth book, and in some ways has been the most difficult to write. Enormous thanks are due to my family, without whose support none of this would be possible. I especially want to thank Kate Nepveu for patient listening, beta reading, and putting up with my odd work schedule and general air of distraction. And thanks to Claire and David for providing great excuses to step away from the computer now and again.

INDEX

ABOUT THE AUTHOR

Chad Orzel is a professor at Union College in Schenectady, NY, and the author of three previous books explaining science for nonscientists: *How to Teach Quantum Physics to Your Dog* (Scribner, 2009) and *How to Teach Relativity to Your Dog* (Basic, 2012), which explain modern physics through imaginary conversations with Emmy, his German shepherd, and *Eureka: Discovering Your Inner Scientist* (Basic, 2014), on the role of scientific thinking in everyday life. He has a BA in Physics from Williams College and a PhD in Chemical Physics from the University of Maryland, College Park, where he did his thesis research on collisions of laser-cooled atoms at the National Institute of Standards and Technology in the lab of Bill Phillips, who shared the 1997 Nobel Prize in physics (not for anything Chad did, but it was a fun time to be in that group). He has been blogging about science since 2002, on his own site, at scienceblogs.com, and most recently for *Forbes*. He lives in Niskayuna, NY, with his wife, Kate Nepveu; their two children; and their new dog, Charlie the pupper.